FIRE DEPARTMENT
INCIDENT SAFETY OFFICER

FIRE DEPARTMENT INCIDENT SAFETY OFFICER

Second Edition

David W. Dodson

DELMAR
CENGAGE Learning™

Australia • Brazil • Japan • Korea • Mexico • Singapore • Spain • United Kingdom • United States

DELMAR
CENGAGE Learning™

Fire Department Incident Safety Officer, Second Edition
David W. Dodson

Vice President, Technology and Trades ABU: David Garza

Director of Learning Solutions: Sandy Clark

Managing Editor: Larry Main

Acquisitions Editor: Alison Pase

Product Manager: Jennifer A. Starr

Marketing Director: Deborah S. Yarnell

Marketing Manager: Erin Coffin

Marketing Coordinator: Patti Garrison

Director of Production: Patty Stephan

Content Project Manager: Jennifer Hanley

Technology Project Manager: Kevin Smith

Editorial Assistant: Maria Conto

Cover Image: David W. Dodson

For product information and technology assistance, contact us at
Cengage Learning Customer & Sales Support, 1-800-354-9706

For permission to use material from this text or product, submit all requests online at **www.cengage.com/permissions**
Further permissions questions can be emailed to
permissionrequest@cengage.com

Library of Congress Control Number: 2007008350

ISBN-13: 978-1-4180-0942-7

ISBN-10: 1-4180-0942-3

Delmar
Executive Woods
5 Maxwell Drive
Clifton Park, NY 12065
USA

Cengage Learning is a leading provider of customized learning solutions with office locations around the globe, including Singapore, the United Kingdom, Australia, Mexico, Brazil, and Japan. Locate your local office at: **international.cengage.com/region**

Cengage Learning products are represented in Canada by Nelson Education, Ltd.

For your lifelong learning solutions, visit **delmar.cengage.com**

Visit our corporate website at **www.cengage.com**

Printed in the U.S.A.
6 7 11

DEDICATION

This book is about pride, professionalism, knowledge, and the effort to protect firefighters. These attributes will ultimately make a difference in the fire service. Therefore, I dedicate this book to the individual who works tirelessly to protect our brother and sister firefighters:

The Fire Department Incident Safety Officer

CONTENTS

PREFACE

The purpose of this book remains the same as that of the first edition: to reduce firefighter injuries and fatalities. With all the efforts and improvements invested in protective equipment, incident management systems, technology, training, and standards, we still have not made a dent in reducing injuries and deaths. Statistics prove this.

In the first edition of this book, I predicted that a well-trained and experienced incident safety officer (ISO) could reduce at least *half* the anticipated deaths and injuries. I further expressed that a dedicated and trained ISO is the quickest, cheapest, and simplest solution to improve on-scene firefighter safety. I was probably overoptimistic, but I remain convinced that a trained and experienced ISO can make a difference. In this book, therefore, I am revising the phrase "trained and experienced" to read "trained, experienced, *and persuasive*."

Since the first edition (1998), I have had the pleasure of watching the role of the ISO mature. Of particular note, the Fire Department Safety Officers Association (FDSOA) realized the goal of creating and implementing an accredited, National Incident Safety Officer Certification. FDSOA should be very proud of this program, and those who have invested the energy to become certified should be equally proud. I have seen hundreds of fire departments implement comprehensive ISO programs, and I have received hundreds (if not thousands) of e-mails describing how an ISO recognized and prevented a sure injury or death.

With the thousands of trained ISOs on the job, why have overall death and injury statistics not been reduced? The answers are unclear, but I may have some of them. First, recent research that tells us that residential and commercial structure fires have changed: They have never been as hot, as fast, or as explosive as they are now. Second, the building that we fight fires in is continually changing. Lightweight constructed buildings *will* sucker punch the firefighter who approaches them with the same comfort as in the 1980s and 1990s. Further, we are responding to more incidents and to a greater variety of incidents. If the fires have changed, if the building has changed, and if the number and type of incidents have changed, then it stands to reason that the ISO must change. That brings us to this second edition.

The first edition of this book focused on selling the ISO concept and giving basic guidance for the performance of ISO duties. This edition still includes some ISO system design suggestions, but it expands the intellectual components to help the ISO better understand incident dynamics and intervene with more effective persuasiveness. In essence, the book is intended to be a training manual and an ISO desk reference.

HOW TO USE THIS BOOK

The book is divided into three sections:

- **Section One** offers an introduction to the ISO role as a way to start preparing for the assignment. General safety concepts, guiding documents, and effective system components are discussed.

- **Section Two** perhaps the most important one, provides in-depth "front-loading" information to help the ISO gain knowledge and skills. Of particular note, an entire chapter is devoted to reading smoke and another to reading buildings. These two areas—more than any—can help the ISO predict hostile events that can trap firefighters.

- **Section Three** shows how the information in Section Two can apply, in a realistic way, to actual incidents. A basic approach to ISO duties is offered, as well as specific considerations for unique incidents, like hazmat and wildland/interface incidents. Section Three ends with the often forgotten ISO postincident responsibilities.

FEATURES OF THIS BOOK

Delmar Learning's hallmark features have added value to the fire service library. You will find that many of Delmar's usable features here will help benefit your learning experience:

- **Street Stories** from actual incidents open each section and provide an inside look at the job of the FDISO.

- **Notes** highlight critical concepts presented in the text.

- **Safety and cautions** outline important safety tips and advice for preventing injuries and deaths on scene.

- **Key Terms and Acronyms** provide learners with the knowledge to effectively communicate on the job.

- **Points to Ponder and Discussion Questions** encourage learners to develop critical-thinking skills for effective response.

- **Extensive photos and graphics** illustrate critical points and procedures explained in the text to enhance learning.

- **Additional Resources** point learners in the direction of references which include further information on specific topics.

- **Review Questions** at the end of each chapter evaluate the learners comprehension of the concepts presented in the chapter.

NEW TO THIS EDITION

The first edition triggered an avalanche of appreciated criticism, corrections, and ideas for a better second edition. Additionally, I have had a 14-year opportunity to meet with hundreds of colleagues while traveling and teaching ISO and Reading Smoke classes all over the country. This edition reflects that input and experience. Specifically, I have added the following:

- An entire chapter on the art of reading smoke for predicting hostile fire events
- A revised, five-step process to help incident safety officers predict building collapse
- Attention to National Incident Management System (NIMS) compliance
- Detailed processes for the expansion of the ISO role at major fires, terrorist events, and disaster incidents
- Specific ISO responsibilities at hazmat, technical rescue, and wildland fires
- Updated material that blends with NFPA 1521, *Fire Department Safety Officer,* 2008 edition (proposed)
- Appendix materials that can help you create checklists and standard operating procedures

SUPPLEMENTS TO THIS BOOK

To further assist instructors, an e.resource CD-ROM is available, containing the following components:

- **Instructor's Guide** with Lesson Plans and Answers to Review Questions
- **PowerPoint**™ presentations containing art from the book
- **Computerized Testbanks** in ExamView
- **Correlation Grid** linking the Second Edition to NFPA 1521, 2007

Order#: 1-4180-0943-1

SUGGESTIONS ENCOURAGED

As author, I take full responsibility for any misinterpretations of the suggestions and ideas that so many have graciously offered. I encourage you to be critical in your use of this edition—envision its application, practice the skills, and provide suggestions and comments so that the third edition will be an even better tool for preventing firefighter injuries and deaths.

David W. Dodson

ABOUT THE AUTHOR

David W. Dodson is a 28-year fire service veteran, having started his fire service Career with the U.S. Air Force and then spending almost seven years as a fire officer and training/safety officer for the Parker Fire District in Parker, Colorado. He became the first career training officer for Loveland Fire & Rescue (Colorado) and rose through the ranks as a company officer, hazmat technician, health and safety officer, duty safety officer, and emergency manager for the city. He accepted a shift battalion chief position for the Eagle River Fire District in Colorado before starting his current company, Response Solutions, which is dedicated to teaching firefighter safety and practical incident handling.

Dave is the author of the first edition of Delmar Learning's *Fire Department Incident Safety Officer,* as well as a coauthor of their *Firefighter's Handbook.* He has also authored many published items on firefighter safety and survival and first-due officer procedures.

Chief Dodson has served as the chairman of the NFPA 1521 Task Group (*Fire Department Safety Officer*) and served on the Fire Service Occupational Safety and Health Technical Committee for NFPA. Dave is also a past president of the Fire Department Safety Officer's Association. In 1997, Dave received international acclaim as the ISFSI George D. Post Fire Instructor of the Year.

Dave is the course developer for the popular Incident Safety Officer Academy and the Art of Reading Smoke programs and has presented programs to over 20,000 fire officers in the United States and Canada.

ACKNOWLEDGMENTS

This book represents my fourth overall textbook adventure. Each time, I am amazed at the amount of energy and synergy that *other* people contributed. I wish I had the space and time to thank everyone involved, and I wish I had the memory cells to remember everyone who planted a seed, gave some insight, or offered a suggestion. To fully appreciate my indebtedness, you must understand that the ideas, experiences, and suggestions in this book come from a host of contributors. I will probably forget someone, but I want to list and offer a colossal thank-you to those who opened a door, provided mentoring, or challenged me to go another step.

In my young lieutenant years, I was lucky enough to work for two fire chiefs who encouraged me to go to the National Fire Academy (NFA) and achieve all I could. I am pretty sure that I broke some rules, but I "exploited" NFA. In doing so, I learned more from hallway gatherings, classroom arguments, and pub BS sessions than any college could ever hope to teach me. Let me thank those two chiefs: Duncan Wilke and Dick Minor. Thank you also to the brothers and sisters who shared what was *really* happening in the fire service as we communed at the NFA.

The NFPA standards process was eye-opening. The NFPA experience taught me that you should not throw darts if you are not willing to be the dartboard. Thanks to all 33 members of the NFPA Fire Service Occupational Safety & Health Tech Committee. A special thank you to Carl Peterson (NFPA Staff Liaison).

The Fire Department Safety Officer's Association (FDSOA) adopted the first edition of this book as the primary study guide reference for National Certification of Incident Safety Officers. Thanks to FDSOA and all those who endeavored to become certified; the feedback from your efforts is included in this edition.

I believe in mentoring. I also believe that most mentors do not know that they are mentoring. A special thank-you goes to Chief Alan Brunacini (retired, Phoenix Fire), Chief Billy Goldfeder (Loveland-Simms, Ohio), and Chief Bobby Halton (editor, *Fire Engineering* magazine). I would like to thank—and remember—Chief Ray Downey and Chief Don Burns (FDNY, 9/11 victims), who shared valuable insight and experiences. Thanks are also due to mentor Francis Brannigan, who left us in 2006; we can never replace him, and his impact on the fire service will live on.

To those who have left distinct fingerprints on this manuscript, I owe more than a thank-you: David Ross (Chief of Safety, Toronto Fire Services), Peter McBride (Safety officer, Ottawa Fire Department), Terry Vavra (Deputy Chief, Lisle-Woodridge FPD, Illinois), John Mittendorf (retired Battalion Chief, Los Angeles City Fire), I. David Daniels (Chief, Renton Fire, Washington), Mark Emery (Battalion Chief, Woodinville, Washington), and attorney David C. Comstock (Chief, Western Reserve Joint Fire District, Ohio). Thanks also to the photographers who provided the visual punch.

Thanks also to those who continually support my efforts: Karen Boor, Rick Davis, Steve Davis, Jeff Edmonds, Mike "Juice" Juozapaitis, Brian Kazmierzak, Carol Lanza, Jeff Money, Katherine Ridenour, Tina Rigitello, and John Tanaka. My dear friends, Scott and Jenny Macumber and Paul and Kim Sparks, have been more than supportive—thanks!

The gang at Delmar Learning are simply amazing: Their support, encouragement, and professionalism have helped make the first and second editions rewarding. Jennifer Starr and Alison Pase are project masters who, with their project team, kept me focused. Join me and the Delmar gang in showing gratitude to the following review group:

Dave Casey
Bureau of Fire Standards & Training
Florida State Fire Marshal
Florida State Fire College
Ocala, FL

Dennis Childress
Orange County Fire Authority
Ranch Santiago College
Orange, CA

Scott Clark
Manhattan Fire Department
Manhattan, KS

Mike Flavin
St. Louis Fire Academy
St. Louis, MO

David Fultz
LSU Fire & Emergency Training Institute
Baton Rouge, LA

Robert Klinoff
Kern County Fire Department
Bakersfield, CA

Jesse Lapin—Bertone
Tamarac Fire Rescue
Tamarac, FL

Rob McLeod III
Chandler Fire Department
Chandler, AZ

Jeff Pindelski
Downers Grove Fire Department
Downers Grove, IL

Richard Powell
Saginaw Township Fire Department
Saginaw, MI

William Shouldis
Philadelphia Fire Department
Philadelphia, PA

Thomas Wutz
New York State Fire Academy
Albany, NY

Finally, and most importantly, I would like to thank my loving family. Kelsie and Dan are two gifted and energetic teenagers who make my life a joy. I wish you both the best as you mature and discover all that life can give—you'll do well! My talented life partner, LaRae, who, more than anyone made this endeavor possible, is an inspiration with her unwavering love, dedication, and positive spirit. I am truly blessed.

PERHAPS A SAFETY OFFICER COULD HELP?

The fire service is a high-stress, active occupation with individuals capable of feats of bravery that many in the general population would not dare to take on. Big fires, breathtaking rescues, and brushes with death itself make the fire service an occupation full of stories that people, both within and outside it, are very interested to hear. Unfortunately, too many such stories end in the death of one or more of the players. The question is how many of these deaths could have been prevented or converted to a survivable injury by the presence of a safety officer?

The safety officer in many fire departments is still perceived as the "safety cop," "wellness zealot," or "fitness freak" who has the unenviable task of convincing tough firefighters that they aren't so tough or the "experienced guys" that their experience won't protect them in certain situations but that working safely will. Though there is a lack of empirical evidence showing how the safety officer's presence positively impacts an incident, there is at least anecdotal evidence that the lack of a safety officer and a specific focus on safety are contributing factors in the unfortunate injuries and untimely deaths of firefighters. True, no one likes the "safety cop" any more than we do a regular cop when we have broken the rules. This is true not only today, but for many years. It was certainly true in my early years in the fire service.

The year 1987 is on record as the seventh deadliest year since the NFPA began tracking firefighter fatalities in 1977. As a young firefighter I had no idea that on a hot July day that year I would witness one of those fatalities, and I had even less an idea or interest in losing another person in a fire station where I was assigned. It had already happened three times in my short career. But unfortunately, on this day it was going to happen—again!

It was early afternoon in my hometown, a place known more for rain than sunshine. We were in the middle of one of those fabulous days that we rarely told people about, perhaps as a subconscious method of keeping migration down. The sun was high in the sky, and it was a lovely afternoon in the city! A rather routine day of basic house duties and nondescript minor alarms was interrupted by the house tones for a "full response" that included at first only the engine company and deputy chief from my station. We sent a number of engines, trucks, and support vehicles. What we did not send or establish was a safety officer.

A safety officer might have questioned the use of opposing tactical operations in the form of exterior large diameter lines in use simultaneously with interior lines. A safety officer might have asked why the many firefighters on the scene were doing things that they thought were worthwhile but that had no connection to the incident commander's (IC's) strategic plan.

Perhaps a safety officer would have convinced the organization that department knowledge of the incident command system would help minimize the possibility of the incident commander getting overwhelmed. Maybe a safety officer would have recognized that a member of my crew had been separated from us for over 45 minutes and might have started a search sooner. Perhaps a safety officer could have encouraged the incident commander to limit all but emergency traffic on the radio once it was determined that someone was missing. A safety officer could have supported the IC's efforts to modify the tactical operations to place the search of the firefighter above the firefight itself, rather than trying to split the on-scene resources to achieve two critical objectives at the same time.

In the end, a member of my engine company died as a direct result of freelancing. He was not the only one who ever went off on his own, just the only one who died doing it. It was a different time and a different day, and what occurred, occurred. Though valiant efforts were made to save the life of the firefighter who died that day, maybe—just maybe—a safety officer could have prevented the chain of events that caused his death. And even if a safety officer couldn't have changed what happened then, maybe it will help tomorrow.

I. David Daniels
Fire Chief
Renton Fire Department, Washington

INTRODUCTION TO THE SAFETY OFFICER ROLE

Learning Objectives

Upon completion of this chapter, you should be able to:

- Describe the emergence of the safety officer role in fire departments.
- Discuss the history of the fire department safety officer.
- List the National Fire Protection Association (NFPA) standards that affect and pertain to the incident safety officer.
- Explain the need for an incident safety officer in empirical and image terms.

THE SAFETY OFFICER: AN INTRODUCTION

safety officer
according to NIMS, a member of the command staff responsible for monitoring and assessing safety hazards or unsafe situations and developing measures for ensuring personnel safety; *note:* NFPA uses the title "incident safety officer" (ISO) for greater specificity

National Fire Protection Association (NFPA)
a for-profit association recognized for developing consensus standards, guides, and codes for a whole realm of fire-related topics

health and safety officer (HSO)
the member of the fire department assigned and authorized by the fire chief as the manager of the occupational safety and health program

The title **"safety officer"** is used daily in fire departments around the country. Often, this title is used to refer to the individual in charge of a department's entire safety and health program. Other times, the title is applied to an arriving engine or truck company officer who reports to the incident commander (IC) and is delegated the safety officer task (**Figure 1-1**). In some departments, the safety officer is actually an OSHA compliance officer. In other departments, the safety officer is another title for the training officer, that is, the person who responsible for ensuring that safety is incorporated into routine, training, and incident activities. Over the past decade, fire departments have discovered that the title "safety officer" is a bit too generic. In the mid-1990s, the **National Fire Protection Association (NFPA)** task group assigned to update NFPA 1521, *Fire Department Safety Officer,* added specificity to the title "safety officer." A **health and safety officer (HSO)** is the person (or persons) assigned and authorized by the fire chief as the manager of the fire department's safety and health program. The duties, qualifications, and authorities of the HSO are spelled out in the NFPA 1521 standard. The NFPA task group further distinguished the safety officer title by adding **incident safety officer (ISO)**, a title given to the person assigned to fill the command staff position responsible for monitoring and assessing safety hazards or unsafe situations and for developing measures to ensure personnel safety at the scene of an incident.

Figure 1-1 *An effective ISO can reduce the chance of firefighter injury or death.*

incident safety officer (ISO)

a member of the command staff responsible for monitoring and assessing safety hazards or unsafe situations and for developing measures to ensure personnel safety at the scene of an incident

National Incident Management System (NIMS)

a presidentially mandated, consistent nationwide approach to prepare, respond, and recover from domestic incidents regardless of cause, size, or complexity

The reason for splitting titles is in recognition of certain realities in the fire service. Small fire departments usually had one person who performed both the HSO and ISO roles. Those departments soon realized that one individual couldn't possibility make every significant incident where an ISO would be desirable and at the same time fill the role of HSO. In large departments, it was clear that the management of a significant occupational health and safety program required knowledge and skills that were markedly different from those required of an ISO. Thus, a division of the safety officer role was written into the NFPA 1521 standard. The HSO/ISO division was first proposed by the Fire Department Safety Officer's Association (FDSOA) when it offered the 1991 class, "Preparing the Fireground Safety Officer." A project team audited this class and went on to help the National Fire Academy develop its own separate HSO and ISO field classes. The NFPA task group recognized these efforts and recommended the division of the HSO/ISO roles in the standard. Interestingly, the **National Incident Management System (NIMS)** still uses the generic title of safety officer for NIMS compliance texts and documents. NIMS was developed through Homeland Security Presidential Directive 5 (HSPD-5) to create and mandate a consistent nationwide approach to prepare, respond, and recover from domestic incidents regardless of cause, size, or complexity. The HSO and ISO titles will be used in this book, although there are "NIMS compliance" discussions suggesting that the ISO title revert to safety officer (SO) and that the HSO title change to health and safety administrator (HSA).

For the sake of clarity, let's take a brief look at some of the roles and responsibilities of the HSO and ISO. As seen in **Figure 1-2,** the HSO is

Health and Safety Officer Functions	Incident Safety Officer Functions
• Risk management planning	• Risk evaluation
• Procedure review	• Resource evaluation
• OSHA compliance	• Hazard identification and communication
• Safety and health education	• Action plan review
• Data analysis	• Safety briefings
• Facility and equipment inspection	• Collapse zoning
• Infection control	• Accident investigation
• Wellness programming	• Postincident analysis
• Accident investigation	• Safety committee participation
• Postincident analysis	
• Safety committee participation	

Figure 1-2 *HSO and ISO functions.*

responsible for health and safety administration, whereas the ISO is focused on scene-specific operations. It is also obvious that some overlap occurs—by design. Overlapping responsibilities provides consistency and communication in the different roles. Before we dive too deep into ISO development, however, let's take a quick journey through the history of today's ISO.

HISTORY

Safety officers, in one form or another, have been present in the American work force for a long time. Some of the first safety officers came out of the fire service. In the late 1800s and early 1900s, "wall watchers" stood at corners of buildings and watched the walls for signs of bowing or sagging during a working fire (**Figure 1-3**). This practice followed the catastrophic collapse of New York's Jenning Building on April 25, 1854.[1] In this tragedy, twenty firefighters were buried following a partial, then significant collapse of the

Figure 1-3   *A late 1800s fire officer— the first safety officer—shouts collapse warnings.*

building at 231 Broadway. In Colorado Springs, in 1898, a decision was made by on-scene officers to withdraw firefighters from a railroad car fire containing black powder. Thirty minutes later, the car detonated, causing a wind-fed fire that destroyed many buildings, including the famous Antlers Hotel. These are just a few examples of the early-day "safety officer" role. In some respects, the fire service was viewed as progressive in the "appointment" of safety officers as part of risk management.

As America became more industrialized, the need for a safety officer increased—for both fire departments and general manufacturing. Signs of this need were felt in World War I when soldiers became mechanized, but it wasn't until World War II that the safety officer role was formalized. It seems that the first industrialized war brought significant injury and death in *support* operations as well as in combat. The military started to look at why people were getting hurt outside of combat, and they appointed safety officers for an immediate impact while they developed safer procedures.

Even as World War II carried on, factories and other manufacturing industries began looking at the safety of their workers, partly because of the inclusion of a significant number of female workers. Some of this introspection came at the request of the insurance industry, while other safety issues came at the request of organized labor. Before long, safety inspections, posters, briefings, and other measures were commonplace in the manufacturing environment. In 1970, Congress passed the William Stieger Act, which included the Occupational Safety and Health Act and created the Occupational Safety and Health Administration (OSHA). President Richard M. Nixon signed the act into law that December. The law gives equal rights and responsibility to employers and employees with respect to safe working conditions. Today, you can find even small businesses with a dedicated safety manager or OSHA compliance officer. In large corporations, an industrial hygienist is tasked to administer risk management and employee safety programs. Safety in corporate America is so prevalent that many careers have been spawned in the safety arena. Colleges, universities, and even vocational schools offer degree and certificate programs in safety and safety-related programs.

FIRE DEPARTMENT SAFETY OFFICER TRENDS

Even though some fire departments have been using safety officers for almost a century, the fire service as a whole was slow to catch on to the concepts of safety and risk management in all phases of fire department operations and administration—at least to the degree established by OSHA. The roots of risk management and a dedicated safety officer in today's fire service lie in the development and 1987 adoption of NFPA 1500, *Fire Department Occupational*

Safety and Health Program. While this standard was slow to catch on, the late 1980s and early 1990s found fire departments trying to integrate the safety officer (risk manager) role into the department culture. From this period, an oft told anecdotal story best symbolizes the incorporation of the Safety Officer. The story goes a lot like this:

> The Fire Chief went to a national conference and heard some guy talk about the need for a department safety officer. NFPA 1500 says you got to have one. The best person for the job is your training officer. Next thing I know, I'm it!

This sounds like a typical fire service story. Unfortunately, not much fire service training material existed to tell the newly appointed safety officer what to do. Most safety officers got a personal copy of NFPA 1500, read it and found that the standard was nothing more than a fire service twist on commonplace practices in the industrial world. Granted, some NFPA 1500 issues were controversial (staffing, equipment design, and the like), but the basic premise to develop and administer an active health and safety program was its guiding purpose.

As a companion document to NFPA 1500, NFPA 1501, *Standard for Fire Department Safety Officer,* was created by the NFPA Fire Department Occupational Health and Safety Technical Committee. This standard addressed the authority, qualifications, and responsibilities of the safety officer, whereas the prior standard addressed primarily the HSO role. NFPA 1501 has since been changed to NFPA 1521 in an effort to standardize NFPA numbering. Both 1500 and 1521 are updated on a regular revision cycle under the guide of NFPA's Fire Department Occupational Health and Safety Technical Committee.

NFPA

NFPA 1501 has since been changed to NFPA 1521 in an effort to standardize NFPA numbering.

Prior to the development of these standards, some safety officer trends were well under way in the fire service. In the 1970s, the FIRESCOPE (Fire Resources of Southern California Organized for Potential Emergencies) program was developed and used for multiagency incidents on the West Coast. A safety officer was listed as a command staff position to help the incident commander with delegated safety duties. In the late 1970s, former Chief Alan Brunacini of the Phoenix Fire Department began teaching a "Fire Ground Command" seminar across the country. In this seminar, it was recommended that a safety officer, or safety sector, be established to provide a higher level of expertise and undivided attention to fireground safety. This sector was designed to report directly to the fire ground commander, as well as to advise and consult with other sector officers. In 1983, IFSTA published "Incident Command System," a manual in which a safety officer position was integral to the command staff; a checklist and organizational chart were included[2] **(Figure 1-4)**.

In other examples, large cities such as New York City were creating safety divisions and shift-assigned safety officers to provide injury investigation

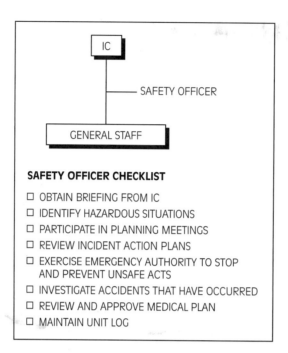

IC

SAFETY OFFICER

GENERAL STAFF

SAFETY OFFICER CHECKLIST

☐ OBTAIN BRIEFING FROM IC
☐ IDENTIFY HAZARDOUS SITUATIONS
☐ PARTICIPATE IN PLANNING MEETINGS
☐ REVIEW INCIDENT ACTION PLANS
☐ EXERCISE EMERGENCY AUTHORITY TO STOP
    AND PREVENT UNSAFE ACTS
☐ INVESTIGATE ACCIDENTS THAT HAVE OCCURRED
☐ REVIEW AND APPROVE MEDICAL PLAN
☐ MAINTAIN UNIT LOG

Figure 1-4 *Initial safety officer checklists were a great start—and overly simple.*

■ **Note**

NIMS is referred to as the "single I" NIMS so as not to be confused with the "double I" NIIMS used by the NWCG.

and incident safety duties. The National Interagency Incident Management System (NIIMS), used by the National Wildfire Coordinating Group (NWCG), recognizes the safety officer as directly reporting to the incident commander. NIIMS is a direct descendant of the FIRESCOPE program. On a humorous note, (NIMS is referred to as the "single I" NIMS so as not to be confused with the "double I" NIIMS used by the NWCG.) In NIIMS, the safety officer is classified as either a Type 1 (SOF1) or Type 2 (SOF2). An SOF1 is qualified to deploy nationwide as part of an incident management team (IMT). An SOF2 is usually a state- or locally qualified safety officer for wildland and interface fires. Interestingly, both the SOF1 and SOF2 must meet the same criteria for qualification.[3]

In 2004, President George W. Bush signed a presidential directive (Homeland Security Presidential Directive 5, *Management of Domestic Incidents*) that mandates the use of NIMS (single "I") as part of the National Response Plan (NRP) administered through the Department of Homeland Security. Within NIMS, the standard title safety officer is used to describe the ISO.

The roles of the HSO and ISO continue to evolve. NFPA seems to be leading the way with the 1521 standard. The assigned 1521 task group is already addressing ISO needs and is proposing further maturity for the ISO role. It is only a matter of time before the National Response Plan incorporates this maturing role in NIMS.

THE NEED FOR AN INCIDENT SAFETY OFFICER

The role of a fire department incident safety officer is based on a simple premise: We (in the fire service) have not done a good job taking care of our own people.

> For 200 years we've been providing a service at the expense of those providing the service.
> —Alan Brunacini, Chief (retired), Phoenix Fire Department

Chief Brunacini was right on target. The 1980s, the 1990s, and now the new century have brought significant improvements in firefighting equipment, standards, and procedures, all with the intent of making the firefighting profession safer. Concurrently, the United States has seen a decline in the number of structure fires. One would think that the combination of better "stuff" and fewer fires would lead to fewer firefighter injuries or deaths. Disturbingly, this is not so. More than ever, the fire service needs to step up its effort to utilize effective ISOs and use them more often. Proof of this need can be found in the empirical and image factors regarding firefighter injury.

Empirical Study

Death and Injury Statistics   About one hundred United States firefighters are killed every year in the line of duty (**Figure 1-5**). On September 11, 2001, 343 FDNY firefighters were killed in the collapse of the World Trade Center Towers. While firefighters will never forget this tragic event, the 343 deaths are often removed from statistical analysis to spot trends more accurately and develop more suitable solutions. Not counting the 9/11 deaths, roughly a third of firefighter fatalities occur at the incident scene. A trained, persuasive incident safety officer can help reduce the potential for these deaths.

Some may argue that this number is acceptable considering the risks that a community expects firefighters to take. However, many fire officers feel that no death is acceptable, especially if it can be prevented by coached aggressiveness. If a single firefighter death can be prevented by the appointment of an incident safety officer, then the effort is worth the cost.

Firefighter injuries (as opposed to deaths) show a slight decline in recent years. This decline, however, falls short of the decline in the number of fires. Perhaps a better measurement is the number of injuries per 10,000 fires fought. This figure shows that the rate of injuries has actually gone up![4] Still, at roughly 50,000 fireground injuries in 2003 (that's 130 per day!), it stands to reason that we in the fire service have a long way to go in injury prevention.

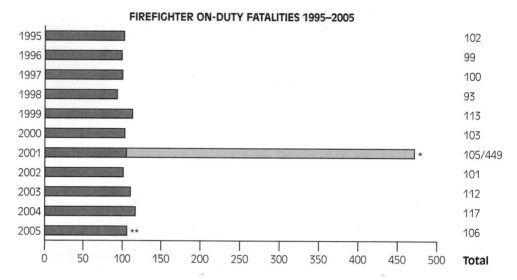

FIREFIGHTER ON-DUTY FATALITIES 1995–2005

Year	Total
1995	102
1996	99
1997	100
1998	93
1999	113
2000	103
2001	105/449
2002	101
2003	112
2004	117
2005	106

*EVENTS RELATED TO THE 9/11 TERRORIST ATTACK (EXCLUDED FOR AVERAGING)
**SUBJECT TO UPDATE/CHANGES

Figure 1-5   *An average of 104 firefighters have died in the line of duty in the past ten years. (Source: United States Fire Administration.)*

Combine firefighter death and injury numbers with a national trend of fewer fires, and one can hypothesize that the effort to reduce firefighter injuries may not be as effective as it could be. It is easy to see that the appointment of an ISO, more often than not, seems prudent. Other programs, such as firefighter wellness and incident management systems, do reduce injury and death potential over time, but the use of an ISO can start to reduce these threat potentials *today*.

How many firefighter injuries and deaths have been prevented by the action of an ISO? We don't know for sure. One Missouri fire chief presented this question when arguing against the cost of formalizing an ISO program for his department during an open forum on safety issues. No one present at the discussion could answer the question with hard data, although many felt that incident safety officers did, in their own experience, change a situation that "could have" led to an injury. In one case, an ISO at an Illinois strip mall fire called for the evacuation of a building, and the incident commander concurred. The firefighters present withdrew and, after protesting the pullout, witnessed the roof slowly collapse into the building. This came just a few short years after a firefighter died in a similar roof collapse in a neighboring department. The point is simple:

We don't keep good data on what could have been, and at times our memories are too short.

Worker's Compensation   When studying the empirical effects of firefighter injury and death, it would be negligent to skip the effects of work-related injuries on worker compensation. Worker's compensation is statutory for each state, and each state's version has its own intricacies. Rates, however, are set by the National Council on Compensation Insurance (NCCI). Each state may also adjust rates for firefighters based on experience in that state; this is called an experience modifier. To determine a worker's compensation rate for a given department in a state, a formula of NCCI rate × payroll × experience modifier is used. The experience modifier is based on a three-year loss experience.

A few points can be made about worker's compensation. First, and most importantly, worker compensation programs are not free; they are costly and the cost is based on history: the number of claims and the cost of the claims. Second, a fire department cannot always shop around for a good rate, as some of us do for automobile insurance. If a firefighter is injured on the job, the ramifications may be felt for many years. Obviously, the more serious the injury is, the longer the impact is felt. Further, this loss impacts all employers with employees in the firefighter class. It is easy to imagine the impact of such injuries on the long-term financial status of a fire department. Creating and funding a total ISO response system may seem expensive, but the costs of *not* funding such a program can be extraordinary if a fire department experiences a firefighter injury or fatality, especially if the loss could have been prevented by an aggressive ISO.

> ■ **Note**
> Worker compensation programs are not free; they are costly and the cost is based on history: the number of claims and the cost of the claims.

Image Study

The image study (how the general public views us and how we perceive ourselves) of firefighter injury and death deals with less tangible results than quantitative data. A firefighter injury requiring hospital care or extended time off creates stress in the workplace. Small career departments have to struggle with finding a replacement, while small volunteer departments have to get by without the injured individual. Large departments have to shuffle people and provide a rover or other fill-in assignment. A less obvious consequence, perhaps, is the firefighter work slowdown following an accident. Most fire officers have seen this reaction: Concern, introspection, and even trepidation fill the fire house following a significant injury. The more serious the injury is, the more pronounced these displays can be. If an investigation follows a serious injury, the workplace effects of the injury can be multiplied to include fingerpointing and taking sides.

Following a firefighter duty-death, labor and management concerns might take the form of private investigations and/or attempts to minimize liability. In some cases, career and volunteer officers have been demoted, suspended,

> ■ **Note**
> Concern, introspection, and even trepidation fill the fire house following a significant injury.

and terminated. These events are often covered in the local media, causing additional department, community, and individual stress.

At the personal level, a firefighter injury can have a damaging effect on the involved families. Stress permeates the injured firefighter's family life. Even the families of other firefighters are affected. Most experienced firefighters have had to justify their involvement with the fire department to a loved one. Usually, they find themselves in this discussion following a news report of a firefighter killed or seriously injured. Many firefighters know of a peer who has resigned, who has been divorced, or who has chosen to enter an assistance program over a comrade's injury or death.

The Bottom Line

■ **Note**
The information in the following chapters can help you tonight, on your next shift, or in making sweeping changes in your department. The material can be that powerful.

The message is simple: The fire service must continue to improve firefighter safety. Obviously we are not where we could be. An incident safety officer can make a difference—*right now*.

It is with that belief that this book is written. The goal of this book is to give you a systematic and meaningful approach to the creation and implementation of an effective incident safety officer program. Further, the intent is to furnish any firefighter with the information necessary to be an effective incident safety officer or for that matter a more safety-conscious incident commander, company officer, or firefighter (**Figure 1-6**). The information in

Figure 1-6 *The creation of an effective incident safety officer program is the incident commander's key to incident safety.*

the following chapters can help you tonight, on your next shift, or in making sweeping changes in your department. The material can be that powerful. The balance of this book is dedicated to the spirit of making a difference.

SUMMARY

Throughout industrialized history, the title "safety officer" has been focused on preventing injury and the loss of life through solid risk management and hazard reduction. In the fire service, the term has been generically applied to persons with administrative as well as incident response duties. Recently, a divergence has taken place and two specialties have emerged: the health and safety officer, which is primarily an administrative or managerial role, and the incident safety officer, which is an incident command staff position.

The strong need for incident safety officers is evident in the on-scene injury and death statistics, which have not dropped dramatically even though fire departments respond to fewer fires. Further, increasing costs associated with firefighter injuries mandate increased prevention measures. The appointment of an incident safety officer on working incidents is a positive step that can help prevent injuries right away.

KEY TERMS

Health and safety officer (HSO)   The member of fire department assigned and authorized by the fire chief as the manager of the occupational safety and health program.

Incident safety officer (ISO)   A member of the command staff responsible for monitoring and assessing safety hazards or unsafe situations and for developing measures to ensure personnel safety at the scene of an incident.

National Fire Protection Association (NFPA)   A for-profit association recognized for developing consensus standards, guides, and codes for a whole realm of fire-related topics.

National Incident Management System (NIMS)   A presidentially mandated, consistent nationwide approach to prepare, respond, and recover from domestic incidents regardless of cause, size, or complexity.

Safety officer   According to NIMS, a member of the command staff responsible for monitoring and assessing safety hazards or unsafe situations and developing measures for ensuring personnel safety. *Note:* NFPA uses the title "incident safety officer" (ISO) for greater specificity.

POINTS TO PONDER

The Need for ISOs

The Fire Department Instructors Conference (FDIC) is one of the world's largest gatherings of firefighters and fire officers. In 2006, over 28,000 attendees converged in Indianapolis, Indiana, to attend workshops, view vendor displays, and network with peers. In one workshop, the leading chiefs of the Phoenix Fire Department shared some interesting news: They mentioned that Phoenix Fire was eliminating the use of an ISO at working incidents. The rationale for this stunning news was that, since an ISO is like a roving incident commander, a single ISO probably cannot make a big difference because he or she cannot be in all places at all times. It was further explained that sector officers should fill the role of ISO in addition to meeting their sector responsibilities. In a follow-up discussion, former Chief Alan Brunacini mentioned that the ISO function should be "embedded" in all sector officer duties. The thinking is that a department can have numerous ISOs if all the fire officers assigned to tactical units view the incident the way an ISO should.

For Discussion:

1. What benefits do you see in having a separate and dedicated ISO?

2. What issues may arise if the ISO function is expected from fire officers assigned to tactical units (sectors, groups, and divisions)?

3. What type of accountability and communication needs have to be in place when the ISO role is decentralized?

REVIEW QUESTIONS

1. What is the difference between an ISO and an HSO?

2. In general terms, explain the history of today's safety officer in the industrial world as well as in the fire service.

3. List and discuss the NFPA standards related to the incident safety officer.

4. What was the significance of the William Stieger Act?

5. How are the monetary costs associated with firefighter injuries and deaths paid for?

6. Discuss current firefighter injury and death trends and the need for incident safety officer response.

ADDITIONAL RESOURCES

National Incident Management System. Washington D.C.: U.S. Department of Homeland Security, March 2004. Available at www.fema.gov/nims. (Note: A revised edition should be available in late 2007).

NFPA 1500, *Standard on Fire Department Occupational Safety and Health Program.* Quincy, MA: National Fire Protection Association, 2007.

NFPA 1521, *Standard on Fire Department Safety Officer.* Quincy, MA: National Fire Protection Association, 2002. (Note: The 2008 edition should be available in late 2007).

NIOSH Firefighter fatality reports: Available at: www.cdc.gov/niosh/firehome.html.

U.S. Fire Administration firefighter injury and death reports. Available at: www.usfa.fema.gov/.

NOTES

1. Paul Robert Lyons, *Fire in America* (Boston: NFPA Publications, 1976).

2. Fire Protection Publications, Oklahoma State University, *Incident Command System* (Stillwater, OK: Author, 1983), 19–20, 61.

3. National Interagency Incident Management System, *Task Book NFES #230* (Boise, ID: National Wildfire Coordinating Group, 1993).

4. The author compared the reduction of structure fires in the past five years to the reduction of structural fire injuries. USFA and NFPA data were used.

Chapter 2

SAFETY CONCEPTS

Learning Objectives

Upon completion of this chapter, you should be able to:

- List the three elements that affect safety in the work environment.
- Discuss the difference between formal and informal processes.
- List the qualities of a well-written procedure or guideline.
- Discuss the external influences that influence safety equipment design and purchase.
- List and discuss the three factors that contribute to a person's ability to act safely.
- Define risk management.
- Identify and explain the five parts of classic risk management.

THEORY VERSUS REALITY: AN INTRODUCTION TO SAFETY CONCEPTS

Lets face it: *Theory can be boring!* Many fire officers would just as soon skip the theory, or book work, necessary to become an incident safety officer, but they crave the practical, challenging, and critical aspects of the assignment. To become an ISO that *can make a difference,* however, fire officers must build a foundation of understanding (that means theory). While most fire officers agree that effective ISOs need a healthy dose of common sense (a sense of reality), they must also be well grounded in recognized safety concepts (theory), which gives them *un*common sense. Uncommon sense can also be defined as the ability of the ISO to ask two questions: What is the worst that can happen here? What is the probability of it happening? To answer both questions, the ISO needs a keen understanding of safety concepts.

To get a handle on safety concepts, the ISO needs to look at the components making up the operational environment: procedures, equipment, and personnel (**Figure 2-1**). To make the operational environment safer, these components should be evaluated with regard to safety and then be coached with solid risk management concepts. Sounds simple? This chapter takes you

■ **Note**

ISOs must also be well grounded in recognized safety concepts (theory), which gives them *un*common sense.

Figure 2-1 *Personnel, procedures, and equipment all play a role in defining safety in operations.*

Figure 2-1
(Continued)

through procedures, equipment, and personnel and concludes with risk management, giving you a solid safety concept foundation.

SAFETY IN THE OPERATIONAL ENVIRONMENT

Procedures

If incident operations are to be safe, each of its three components needs to be addressed. The term "procedure" is used in a very generic form here to describe all sorts of formal (written) and informal processes that are in place in a fire department. Procedures or processes make up the structure for all the activity at an incident. The first-arriving engine at a fire alarm activation is the most likely to start a series of processes to investigate the alarm.

Formal processes are in writing and they take on many forms: standard operating procedures (SOPs), standard operating guidelines (SOGs), departmental directives, temporary memorandums, and the like. In some departments, formal processes are derived from standard evolutions or lesson plans. These evolutions and lessons can be drilled periodically, on a rotating basis, to ensure that a crew's response to a given situation is appropriate. Some departments adopt training manuals as their operating standard. When a manual gives choices, for instance a hose load, the chosen load can just be circled in the manual. No matter what the source, the key component is that formal processes and evolutions are in writing. In taking this approach, the department achieves consistency in its operations.

procedures
strict processes with little or no flexibility

guidelines
adaptable templates that give wide application flexibility

Many departments around the country have adopted SOGs, in lieu of SOPs, in the belief that a guideline is more flexible and therefore more usable by line officers and incident commanders. A recommendation to use this vernacular term was made to the Lisle-Woodridge Fire Protection District (IL) by its insurance carrier as part of a scheduled audit. Some fire departments recognize both procedures *and* guidelines. In this context, **procedures** are strict processes with little or no flexibility, and **guidelines** are adaptable templates that give wide application flexibility.

Informal processes are processes and operations that are obviously part of a department's routine but are not written. Informal processes are passed on through new member training as well as in day-to-day operations. An example of an informal process is the practice of having firefighters place a full-face hood across their bunker boots so that they are reminded to don the hood before placing their feet in the boots. Another example is the apparatus operator's use of a grease pencil to make status marks on pump panel gauges as a quick reference; at a glance the operator can see if changes have taken place.

Both formal and informal processes can increase the overall safety of a department.

The obvious first step in the development of a formal SOP or SOG (we use "SOP" for the remainder of this chapter) is establishing an administrative process to create, edit, alter, or delete established processes. Once the process is in place, a general format for SOP appearance and indexing is necessary. As seen in **Figure 2-2**, the department has chosen to classify SOPs by topical areas. **Figure 2-3** shows a typical SOP format.

Once topic areas have been defined, the writing of SOPs can begin. It makes sense to write the most important ones first, but which are the most important? The department can approach this question one of two ways, and either can be effective. One way is to perform a needs assessment and flag the areas where line firefighters and officers need guidance. The other way is to look at external influences such as OSHA regulations, insurance services office rating schedules, NFPA standards, and other requirements and determine which areas impact the department most by *not* having a related SOP. Departments choosing the latter route find that items such as personal protective equipment, self-contained breathing apparatus (SCBA), equipment maintenance, and patient care get high priority in the writing effort. As a starting place, SOPs should exist for:

- Use of PPE and SCBA
- Care and maintenance of PPE and SCBA
- Risk/benefit principles
- Emergency driving
- Highway and traffic safety at incidents
- Accident/injury procedures and reporting
- Incident scene accountability

1. 1.1 Incident Command System
2. Emergency Ground Operations
    2.1 Rapid Intervention Company (RIC)
    2.2 Gas/odor investigation*
    2.3 Auto alarms*
    2.4 Train fires*
    2.5 Vehicle fire*
    2.6 Fires at postal facilities*
    2.7 Emergent driving procedure
    2.8 Kaneb pipeline response
    2.9 Volunteer and fire apparatus placement for motor vehicle accidents (MVAs)
    2.10 Operations involving Thompson Valley ambulance
    2.11 Minimal staffing for interior firefighting*
    2.12 Fire ground formation and activation of companies*
    2.13 Standard fire attack procedures/dwelling fires*
3. Alarm Levels/Dispatching
    3.1 City alarm level assignments
    3.2 Rural alarm level assignments
    3.3 Fire resource officer*
    3.4 Fire alarm panel operation and response policy
    3.5 Mutual/automatic aid agreement*
    3.6 Staffing considerations during adverse weather conditions
    3.7 Cancellation procedures for emergency medical service (EMS) and MVA incidents
4. Hazardous Materials
    4.1 Hazardous materials operations*
5. Emergency Medical Services
    5.1 Duties for non-EMS certified personnel
6. Aircraft Rescue and Firefighting (ARFF)
    6.1 ARFF standby policy
7. Technical Rescue and Special Operations
    7.1 Vehicle extrication
    7.2 Rope
    7.3 Trench*
    7.4 Collapse rescue*
    7.5 Confined space*
    7.6 Farm equipment and industrial rescue*
    7.7 Loveland dive rescue standard operating procedures
    7.8 Use of Civil Air Patrol
*These policies still need to be developed and/or approved.

Figure 2-2 *A sample SOP index.*

Purpose:

To establish policy and direction to all department members regarding minimal staffing and resource allocation for safe and aggressive interior structural firefighting.

Responsibility:

It is the responsibility of all officers and firefighters engaged in firefighting operations to adhere to this policy. The Incident Commander is accountable for procedure included within this policy.

Procedure:

1. This policy is applicable to situations where the Incident Commander (IC) has made a tactical decision to initiate an *offensive fire attack*, by firefighters, inside the structure. Additionally, tactical firefighting assignments that expose firefighters to an atmosphere that is *immediately dangerous to life and health* (IDLH) dictate the application of this policy.[1]

2. Prior to initiating interior fire attack or exposure of firefighters to an IDLH atmosphere, a **minimum of four (4) firefighters shall assemble on scene.**[2] These four members shall utilize a "two-in, two-out" concept.

3. The *"two-in"* firefighters that enter the IDLH atmosphere shall remain as partners in close proximity to each other, generally fulfilling the operational role as the *FIRE ATTACK GROUP.* As a minimum, the *"two-in"* firefighters entering the IDLH atmosphere shall have full PPE, with SCBA and PASS device engaged, and have among them a two-way portable radio, forcible entry tool, and flashlight or lantern.

[1] An *IDLH atmosphere* can be defined as an atmosphere that would cause immediate health risks to a person who did not have *Personal Protective Equipment (PPE)* and/or *Self-Contained Breathing Apparatus (SCBA).* This includes smoke, fire gases, oxygen deficient atmospheres, or hazardous materials environments. For Loveland Fire and Rescue application, an IDLH atmosphere can be further defined as an environment that is *suspected* to be IDLH, has been *confirmed* to be IDLH, or *may rapidly become* IDLH. The use of full protective equipment including an activated SCBA and an armed PASS device is mandatory for anyone working in or near an IDLH atmosphere.

[2] The firefighters must be SCBA qualified and capable of operating inside fire buildings without immediate supervision.

Figure 2-3   *A sample SOP format.*

- Trapped and/or lost Mayday procedures
- Emergency evacuation at incidents
- Use of the incident command system
- Effective incident rehabilitation for responders
- Infection and exposure control
- Employee right-to-know (hazards of firefighting)

What makes a good SOP? The answer is simple: Firefighters follow it! This is easier said than done. Good SOPs start with good writing. Good writing starts with a clear outline and the use of simple language.

The outline can come from an officer's meeting, direction from the chief, or a sample from another department. Using the format in Figure 2-3, the author should address the reason (purpose) for the SOP, followed by the responsibility of each affected member for the SOP. Some SOPs have responsibilities at different levels. For example, firefighters may have the responsibility to insure their accountability name tags have been placed on the company's passport. The company officer, on the other hand, may have an oversight responsibility to make sure that all crew members are represented on the passport and to process the passport based on his or her company assignment. The department would have to make sure that a usable policy exists for accountability and training is provided for system use. Other qualities of a good SOP include:

- Simple language
- Clear direction
- Tested technique
- Easy interpretation
- Applicability to many scenarios
- Specific only on critical or life-endangering points

The benefits of clear, concise, and practiced SOPs are numerous. They can become a training outline, a tool to minimize liability, and certainly a tool to guide your members. Above all, a well applied SOP improves departmental *safety*!

The ISO's role in procedures deals with application and review, something like a quality control officer's function. To be effective, the ISO needs to know which SOPs are being applied to a given situation and whether the SOP is accomplishing what is intended. When the SOP is not being used appropriately, or when it exists but is not being used, the ISO needs to interpret whether the actions of firefighters meet the intent of the SOP or if an injury potential exists because the SOP is not being followed. The practical application of SOPs puts the ISO in the *best* place to suggest changes to SOPs or even help create new ones for the department. The ISO who witnesses a failure to follow SOPs during an incident should make a notation and bring up the infraction during postincident analysis or the next scheduled safety committee meeting. If the failure to follow a SOP presents a potential or imminent danger, the ISO must intervene to prevent an injury.

Equipment

In the past few years the fire service has seen a veritable explosion in new equipment designed uniquely for improved safety. What works? What doesn't?

Safety
● Above all, a well applied SOP improves departmental *safety!*

■ Note
The ISO's role in procedures deals with application and review, something like a quality control officer's function.

■ Note
The ISO who witnesses a failure to follow SOPs during an incident should make a notation and bring up the infraction during postincident analysis or the next scheduled safety committee meeting.

What's a fad? What's here to stay? What is fluff? What is essential? How much does the equipment cost? Is it worth it? How long will it last? Will it be outdated soon?

With so many questions and so much time spent answering the questions, too often fire department efforts to improve firefighter safety become focused on equipment, and there is a tendency to blame equipment following an accident. (We have all seen how much easier it is to blame equipment than to blame a person.) To some degree, this blame is predictable. Let's call it the Blame Game.

Blame Game 1:

Chief, it wasn't my fault, the darn [*insert name of equipment here*] broke.

Or:

Chief, if I only had one of those new [*insert name of equipment here*], this would have never happened.

■ **Note**
Equipment helps, but it is arguably the least important factor in the operational triad of procedure, equipment, and personnel.

Equipment helps, but it is arguably the least important factor in the operational triad of procedure, equipment, and personnel. Yet when building an understanding of the safety concept, we have to explore equipment and how it can improve a department's safety. The following factors can be used to evaluate equipment, its selection, and its use.

Department Mission   By looking at a fire department's scope of offered services, we can quickly determine whether it lacks the equipment necessary for safe operations. This is actually quite easy to accomplish. To start, department officers should meet and make a list of the types of incidents handled by their jurisdiction. This is accompanied by a corresponding list of equipment necessary to *safely* handle the incidents (to the degree that the fire department is responsible). As an example, many departments faced an influx of service calls for the activation of residential carbon monoxide (CO) detectors, which are designed to activate with as little as 20 parts per million of CO present in the air. Yet many fire and rescue agencies lacked calibrated instrumentation to confirm the presence of CO in a home. From this national experience, many departments began carrying high-tech, multigas monitors to assist in the safe handling of this type of incident.

With the two lists in hand, officers must discuss the equipment possibilities and place a check mark next to the items that are *essential* to safe operation and a circle next to the *nice-to-have* items. They have to stay focused in this process. While policy and procedure, as well as training needs, are important components, they may distract the process. The officers then compare the list of required equipment to the equipment on hand. Items that need to be obtained can then be prioritized for budgeting and appropriation.

External Influences   When looking for equipment to make incident operations safer, officers need look only to the advertising pages of the many trade journals or scan through the dozens of safety supply catalogs sent to the firehouse. A better tack, however, is to look at *required* equipment. Although requirements vary from state to state, you can look to the following for help on what is required:

- *OSHA regulations.* Known as CFRs (Code of Federal Regulations), these codes often outline the equipment required for a given process to be accomplished. At this writing, states covered under a state-sponsored OSHA plan may have more stringent equipment requirements for public agencies. Those without a plan do not require OSHA compliance from public agencies. For example, Colorado has no state plan; fire departments have no obligation to follow OSHA. Washington State, however, has its own plan (Department of Labor and Industries) and compliance is mandatory for all public agencies. OSHA reform is constantly being debated at the federal level. Soon, all public agencies may fall under more restrictive federal OSHA regulations.

- *NFPA standards.* The vast majority of fire service equipment is tailored to meet or exceed NFPA standards. These consensus standards are designed to offer a minimum acceptable standard for equipment design, application, and maintenance.

- *NIOSH, ANSI, and UL.* Many equipment manufacturers use these agencies to show that their equipment meets or exceeds design and performance requirements.

Equipment Maintenance   As most firefighters know, equipment utilized for incident operations is no better than the care and maintenance it receives. Following an injury accident, much time is spent evaluating the performance of involved equipment. Often, the piece of equipment is inappropriate for the application or not operationally sound.

Because many firefighters may use and maintain a piece of equipment, the complete documentation of repairs and maintenance is essential. Further, a complete set of guidelines should be developed or adopted for essential equipment. Rich Duffy, Director of Occupational Health and Safety for the International Association of Firefighters (IAFF), and Chuck Soros, retired Chief of Safety for Seattle, Washington, suggests considering seven items when writing equipment guidelines:[1]

1. Selection

2. Use

3. Cleaning and decontamination

4. Storage

5. Inspection

6. Repairs

7. Criteria for retirement

The Right Equipment   A quick look at firefighter injury and death statistics shows *what* equipment can make a difference. The following are equipment items that have made a difference in firefighter safety over the past few years. This list is designed to stimulate conversation in your department, hopefully leading to wise equipment changes or purchases. Like any equipment, the following equipment is worthless if it is not used and maintained by trained firefighters.

Personal Protective Equipment (Figure 2-4):

- Accountability passports
- Disposable EMS masks/gloves
- Water-free hand disinfectant
- "High visibility" materials/colors

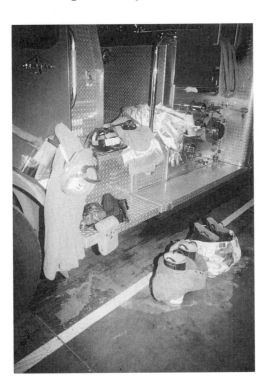

Figure 2-4
Firefighters must now choose from multiple types of protective clothing.

Figure 2-5 *Fire apparatus continues to evolve with added safety features.*

- Nomex/PBI/P84/Kevlar® materials
- Integrated personal alert safety system (PASS) devices

Apparatus (Figure 2-5):
- Enclosed cabs
- Intercom/radio headsets
- Three-point seat belts for all riding positions
- Quick-deploy scene lighting
- Mobile data terminals (laptop computers)
- Ergonomically friendly hose beds
- Vertical exhaust pipes
- Wide reflective trim
- Roll-up compartment doors and roll-out trays
- Global positioning systems (GPS)
- Automatic vehicle locators (AVL)

Tools (Figure 2-6):
- Multigas detectors/monitors
- Speed-shores
- Rehabilitation kits

Figure 2-6 *High-tech tools allow firefighters to work more safely and monitor their health.*

- Command/accountability status boards
- IMS position identification vests
- Two-way radios for each firefighter on a crew
- Thermal imaging cameras (TICs)

Station Equipment (Figure 2-7):
- Exhaust removal systems
- Aerobic and strength exercise devices

Figure 2-7 *Physical fitness equipment is actually "firefighter safety" equipment.*

thermal protective performance (TPP)
a value rating given to the insulative quality of structural personal protective clothing and equipment

- Dedicated disinfection systems/areas
- Fire suppression sprinkler systems
- Extractors for washing structural firefighter clothing
- Open-air/forced-air protective gear storage systems

The effective ISO understands the relationship of equipment to safety. Remember, equipment is arguably the least important facet of the safety triad. In some cases, equipment designed to improve safety can actually lead to greater risk taking. Take the structural fire suppression ensemble as an example. The insulative quality of structural gear is given as a relative value known as the **thermal protective performance** (TPP) rating. A TPP rating is quite scientific, although in simple terms it is a measurement given to the durability of equipment when exposed to a flash fire event. Today's gear has such high insulative qualities (TPP) that it could mask the sensation of heat and allow a firefighter to move into a dangerously hot environment. The effective ISO understands this.

Personnel

When discussing the effect of people on safety, many opinions, philosophies, and emotions have to be considered. It is easy to blame a safety deficiency on equipment or on a poor or nonexistent procedure. Once again, we can see this in our day-to-day station dialogue:

Blame Game #2:

Chief, if we only had a procedure that addressed [*add issue here*], this would have never happened.

Or:

Chief, I'm really sorry Firefighter Jones got hurt. But I followed SOP #302 exactly.

It is more difficult to address the "people" component of the safety triad because of the opinions and emotions involved. Regardless, personnel are an essential factor in improving safety.

> Three factors contribute to a person's ability to act safely: training, health, and attitude.

Training   A successful safety program usually works in tandem with a successful training program. Conversely, an organization plagued by injuries or suffering from costly accidents usually has a deficiency in its training effort.

As it relates to safety, what makes a training program effective? First, some specific qualities should be present in the training:

- Clear objectives
- Applicability to incident handling
- Established proficiency level
- Identification of potential hazards
- Definition of the acceptable risk to be taken
- List of options, should something go wrong
- Accountability to act as trained

Second, the training program must include the right subjects. Although arguments can be made for which training subjects or behaviors are most important for safe operations, a compelling list can be developed based on firefighter injury and death statistics. **Figure 2-8** is a list of training subjects that directly affect incident safety; if personnel are trained in these subject areas and if they appropriately apply the training, incident operations will become safer. For each item in the list, there is an expectation of the depth of understanding and methods needed.

Health The safety and well-being of firefighters increase with their health. Much has been written on the benefits of healthy firefighters, most of which centers on *physical* health. Yet stress or overexertion continues to lead in causes of firefighter duty-deaths and is a significant contributor in injuries. To handle the inherent stress of firefighting, each firefighter's body must be accustomed to and capable of handling stress. Additionally, firefighters need to protect themselves from, and prevent the spread of, communicable diseases

▌ Safety
●
Yet stress or overexertion continues to lead in causes of firefighter duty-deaths and is a significant contributor in injuries.

Essential Training Subjects for Increased Incident Safety	
Subject	**Degree of Understanding**
• Personal protective equipment	Mastery
• Accountability systems	Mastery
• Company formation and team continuity	Mastery
• Fire behavior and phenomena	Proficient
• Incident command systems	Proficient
• Apparatus driving	Proficient under stress
• Fitness and rehabilitation	Practitioner

Figure 2-8 *Injury and death statistics suggest that essential training subjects be addressed.*

and infections. The following are some keys to improving physical health and therefore department safety.

- Annual health screening for all firefighters and line officers
- Vaccination and immunization offerings
- Employee assistance programs (EAP) for alcohol and drug dependencies as well as workplace stress
- A process to determine fitness for firefighting
- Work hardening and mandatory ongoing fitness programs
- Firefighter-fueling (nutrition) education
- Effective rehabilitation strategies, including hydration, active cooling, and refueling

Attention to *physical health* is indeed important. *Mental health* is also important to firefighter safety. Thanks to the recent attention to critical incident stress debriefing (CISD), fire departments are becoming more concerned with the mental health of firefighters. The following are keys to supporting their mental health.

- Training and understanding of critical incident stress signs and symptoms
- Creating a process to have a CISD team activated or available for unusual events or at the request of one or more responders
- Including the firefighter's family in department events
- Making available an EAP for job stress coaching and work or family issues

Attitude   Of all the people factors affecting safety, attitude is the hardest to address. Perhaps this is why firefighter attitudes receive the least amount of attention in terms of safety. People tend to compound safety problems by placing blame after an accident.

Blame Game #3:

Chief, it's not my fault, how was I to know [*fill in someone's name*] was going to[*do whatever*].

Or:

Chief, I was just doing what we've always done. I didn't think [*fill in the event*] could ever happen to us.

Many factors affect the attitude of an individual, and attitudes are dynamic. Of the many factors affecting safety attitudes, the following few are especially prevalent in the fire service:

The department's safety *culture* is made up of the ideas, skills, and customs that are passed from one "generation" to another. How does one see and, more importantly, measure a department's safety culture? One way to illustrate a safety culture is with two conversations at the firehouse.

Sometown Fire Station 1:

Apparatus Operator: Hey Lieu', I'm not going to put the cover back on the grease pit. Jim's bringing down Engine 2 a little later.

Lieutenant: Hope none of the visitors fall in.

Apparatus Operator: Well, there's a big yellow stripe around it.

Lieutenant: OK, but I didn't see anything.

Anytown Fire Station 1:

Apparatus Operator: Hey Lieu', I'm not going to put the cover back on the grease pit. Jim's bringing Engine 2 down a little later.

Lieutenant: That may not be a good idea. The station is wide open and I'd hate to see anyone fall or trip into that pit.

Apparatus Operator: Well, there's a big yellow stripe around it.

Lieutenant: I know, and I know the cover is heavy and hard to move back and forth . . . but I'm serious, I don't want anyone to run into that hole. Why don't you put some traffic cones around it?

Apparatus Operator: That'll work. [*Walks away.*]

It is easy to see the two different attitudes toward safety. The culture of the department may be reflected in its daily conversations or in its actions. In a unique example, the Denver Fire Department experienced a significant accident in which two apparatus collided en route to a reported fire. The department ruled that the most significant factor leading to the accident was the department's attitude that condoned competition between companies to be the first one to get water on the fire.

Fire departments with a long and proud history of no duty-fatalities or significant injuries can fall into another trap: Simply put, they believe such events cannot happen in their department. I've heard this numerous times. The conversation goes something like this: "Did you hear about the firefighter death in [*state name*]. Yeah, the firefighter got trapped in a flashover. They must not 'get it.'" While the opinion may be on target, the underlying message sets a trap.

The department's firefighter *death or injury history* is a factor. A firefighter duty-death often shocks a department's members into an attitude change. Dr. Morris Massey calls this a "significant emotional event" in his renowned videotape, *You Are What You Were When.* A traumatic death is capable of changing a person's value programming, often in the direction of a more healthy safety attitude. While some departments may dismiss a death as a pure accident, most seek to change the way they do business to ensure that the event never repeats itself. The death of Bret Tarver at the Southwest Supermarket fire in Phoenix (March 14, 2001) triggered a sweeping change in procedures, training, and attitudes.[2] This is one of many examples of how a significant emotional event can initiate change.

The *example (or lack of it) set by the line officers and veteran firefighters* is very important. Legendary Notre Dame football coach Knute Rockne once said, "One man practicing sportsmanship is far better than a hundred teaching it." The same can be said about safety. Is safety merely being taught, or is it being practiced? One look at your own department can show whether the following safety indicators are being practiced.

- *Crews or company members are watching not only themselves, but also their team members.* Some examples are crews that make quick, head-to-toe checks of each other just prior to an interior firefighting entry; crew integrity that prevails at *all* incidents *all* the time; company officers who give brief safety reminders prior to tactical assignments; tool operators who voluntarily pass a tool to another operator when initial efforts are unsuccessful; firefighters who offer protective equipment reminders that are welcome and expected; REHAB and SCBA attendants who are organized for quick recognition of fatigue and equipment problems.

- *Work areas are neat and organized.* A Pennsylvania safety officer once said that he could tell if a department had embraced safety from a simple tour of the apparatus bay and the firehouse lounge. Although the evidence is clear that a clean workplace is a safe workplace, it is best to look to the actions of individuals. Do firefighters routinely correct trip hazards while working on a project? Are swing-open compartment doors closed as soon as a tool is retrieved? Are doorways kept clear at the station as well as at the incident site? Do apparatus operators routinely point out obstacles to masked-up firefighters? Does out-of-service equipment get immediate flagging at the incident scene?

- *Drivers are calm, consistent, and attentive.* Safe drivers are usually the ones who follow a simple routine that begins with a confirmation of the incident location. This is followed by communication with the company officer about location, route, or traffic concerns. The driver proceeds to the apparatus in such a way as to get a 360-degree or three-side view of the apparatus. The driver does start-up and belt checks, and then a passenger check (is everyone belted in?). After a go signal, he or she makes a mirror check, looks up at the bay door, and visually scans the apron. Finally the vehicle moves. The driver stops before entering the roadway. Out on the road, the driver gives the sense of control with very few quick-jerk movements: Acceleration is smooth, braking is firm and straight, cornering is like riding a rail. The driver's eyes are always moving and attentive. Face muscles are relaxed, and both hands are graceful in steering and shifting.

- *Observations are openly shared.* Incident safety officers see one of the most reassuring measures of instilled safety values when firefighting teams and company officers report hazards to *them.* Another positive indicator is when personnel are spending time looking *up* and looking *around.* Teams are pointing at walls, wires, and windows. Among the crews is heard, "Watch out for this . . ." or "Keep an eye on that." The crews themselves will put up exclusionary barrier tape around firefighter hazards or collapse zones. The more you see and hear of these behaviors, the further advanced are the safety values of the firefighters.

■ Note
Attitude changes are slow and often emotional, and they require lots of buy-in.

Presumably you can look at this list and see where your department stands on the attitude scale. Remember, however, that attitude changes are slow and often emotional, and they require lots of buy-in. Set personal goals for yourself. Be the example, and then try working for small but steady changes in the department. Be patient.

RISK MANAGEMENT

risk
the chance of damage, injury, or loss

risk management
the process of minimizing the chance, degree, or probability of damage, loss, or injury

Each and every day we take risks. **Risk** can simply be defined as the chance of damage, injury, or loss. **Risk management** is the process of minimizing the chance, degree, or probability of damage, loss, or injury. Most risk managers use a five-step process called *classic risk management* (**Figure 2-9**). An understanding of this process can help the incident safety officer make a difference.

Figure 2-9 *The five-step risk management model is used by risk managers worldwide.*

Five-Step Risk Management
1. Identify hazards
2. Evaluate hazards
3. Prioritize hazards
4. Control hazards
5. Monitor hazards

Five-Step Risk Management

Step 1: Hazard Identification   This is the primary function of an incident safety officer. In the fire service, we may view many operations as routine and not as hazardous, but they are dangerous nonetheless. A great example is smoke. A firefighter breathing through an SCBA does not see smoke as a hazard, whereas the unprotected civilian avoids the smoke to prevent coughing and tear-filled eyes. With today's plastics, one breath of dark smoke can cause dizziness, loss of sensation, and even unconsciousness. Benzene, a known carcinogen, can cause lung cancer with one exposure. Hydrogen cyanide is more prevelant today and can linger long into overhaul operations. Yet how often do you see firefighters breathing smoke? Identifying the hazard is the first step.

Step 2: Hazard Evaluation   Once a hazard has been identified, it has to be assigned relative importance. We call this "hazard evaluation" in the five-step process. In this step, a value is established for a hazard in terms of frequency and severity. *Frequency* is the probability that an injurious event can happen, and it can best be described as low, moderate, or high based on the number of times that a particular hazard is present or the number of times an injury results from the hazard. The same descriptions can be applied to severity. *Severity* can be viewed as harmful consequence or cost associated with injury or damage from a given hazard. With this approach, one can see that any hazard falls into one of the nine categories indicated in **Figure 2-10.**

With this matrix, a value can be assigned to a given hazard. This value helps to determine the priority, or level of importance, of the hazard. This leads us to the next step.

Hazard Evaluation Matrix

		Frequency		
		High	**Moderate**	**Low**
S e v e r i t y	**High**	High/High	High/Moderate	High/Low
	Moderate	Moderate/High	Moderate/Moderate	Moderate/Low
	Low	Low/High	Low/Moderate	Low/Low

Figure 2-10   *A recognized hazard should be placed in one of these boxes based on the potential severity and frequency of the hazard.*

Step 3: Hazard Prioritization   Obviously, a hazard that ranks as high frequency/high severity is one we want to avoid or immediately correct at all cost. Conversely, a low frequency/low severity hazard does not warrant immediate attention. A good example is the classic division of fireground strategies: offensive and defensive. A well involved fire that has captured the attic space in a lightweight wood construction is a high frequency/high severity situation, that is, one that will produce a devastating collapse (and potentially severe injury) in virtually every case. On the other hand, we do not spend much time worrying about the hazards associated with investigating smoke caused by overcooked popcorn in a microwave. One method to simplify this matrix and priority system is to divide the matrix into three hazard classes, as indicated in **Figure 2-11:** priorities 1, 2, and 3.

As a starting place, the ISO should address any hazard that falls in the priority 1 category. During some incidents, the ISO may never get an opportunity to address priority 3 items. If the incident is such that only priority 1 hazards get attention, then the ISO or Incident Commander may consider expanding the safety role to include ASOs or change the incident action plan to better fit the hazards present.

Step 4: Hazard Control   Once the hazard has been prioritized, efforts can be made to minimize exposure or to correct the hazard. Hazard control methods include avoidance, hazard transfer, and hazard adaptation. For firefighting operations, hazard avoidance and transfer are not always possible. Hazard adaptation, however, is the control method most often employed on an incident scene. Hazard adaptation can be accomplished many ways and in

Hazard Priorities

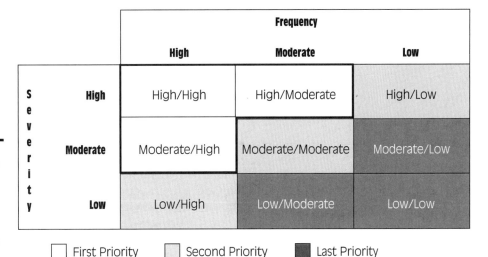

Figure 2-11 *Once a hazard is classified in one of these boxes, a priority can be assigned to it. This helps the ISO juggle multiple hazards.*

many forms. All adaptation methods are designed to make the hazard less severe for the exposed firefighter. This is called *mitigation,* which is usually accomplished by means of ten classic countermeasures:

1. Prevent the *creation* of the hazard.
2. Reduce the *extent* of the hazard.
3. Prevent the *release* of the hazard.
4. Modify the *rate of release* of the hazard.
5. *Separate* the hazard by *time and space.*
6. *Separate* the hazard by a *barrier.*
7. Modify the *basic quality* of the hazard.
8. Make the hazard *resistant* to injury.
9. *Counter* the damage done by the hazard.
10. *Stabilize/repair* the damage done by the hazard.

Step 5: Monitoring Hazards   If the risk management approach is effective, the department should see a decline in injuries, accidents, and close calls over time. Changes in equipment, in staffing, in procedures, and in general can create, alter, or eliminate hazards. Constant monitoring can catch the changes and lead to proactive hazard control. In one example, a city government experienced a notable increase in employee back injuries. The city risk manager hired a noted back injury prevention specialist and in less than two years virtually eliminated back injuries. The program paid for itself in those two years by a reduction in worker compensation claims. This, in turn, lowered the city's loss history, which lowered its annual premium. The savings were used to fund employee benefits (health club memberships).

At an incident, the ISO is always monitoring hazards, even after hazard countermeasures are implemented. This is cyclic thinking, that is, the ability to revisit hazards and continually weigh the operations and the environment to see if a hazard is truly being mitigated. Just as a fire is dynamic, so must ISOs be cyclic in their evaluation of risk. An excellent phrase that captures the essence of this last step was presented in the U.S. Fire Administration's publication, *Risk Management Practices in the Fire Service:* "Risk Management is a system, not a solution."[3]

Risk/Benefit Thinking

The five-step risk management model is a process for addressing hazards. Practitioners of the model often overlay the specifics of the five steps with a simple question: Are the risks being taken by people (the exposure to potential and present hazards) worth the benefit that can be gained? We call this risk/benefit thinking. The hallmark of a good ISO—and any decision maker for that matter—is the ability to continually reassess risk versus benefit. We talk about this more in depth in Chapter 8.

SUMMARY

To be effective, the incident safety officer must have a solid foundation in general safety concepts and risk management. The ISO must appreciate the roles of workplace procedures, equipment, and personnel, as well as how to achieve a safe workplace through the evaluation of and improvement in these areas. In many cases, improvement may be difficult, especially in the area of attitudes. Risk management is the process of minimizing the chance for an injury or loss or minimizing the degree of injury or loss. The most common approach to risk management is the five-step classic risk management model. The incident safety officer can apply the concepts of classic risk management and actually prioritize hazards that need attention through countermeasures. Continually monitoring risks and applying risk/benefit thinking are also necessary to complete the task of making a difference.

KEY TERMS

Guidelines   Adaptable templates that give wide application flexibility.

Procedures   Strict processes with little or no flexibility.

Risk   The chance of damage, injury, or loss.

Risk management   The process of minimizing the chance, degree, or probability of damage, loss, or injury.

Thermal protective performance (TPP)   A value rating given to the insulative quality of structural personal protective clothing and equipment.

POINTS TO PONDER

Safety Concepts

The death of a fellow firefighter conducting fire suppression activities usually gives us a chance to pause and reflect on our own mortality. Most of us want to know the circumstances of the event so that we can minimize the risk of a similar occurrence—or just outright prevent it from happening to us. In February 2005, a fire captain with eleven years of career service (three as captain) died after being trapped by the partial collapse of the roof of a vacant one-story wood frame dwelling in Texas. The house was abandoned and in obvious disrepair. Area residents referred to the building as a "crack house." The 1950s ranch-style house was quite small and had two additions built onto the back of it sometime after its initial construction. Arriving firefighters reported fire venting through the roof at the rear of the house. The captain and a firefighter entered the front of the house as part of the initial attack. Visibility was reported to be good in the front of the house but changed quickly as they advanced toward the rear. The crew had just started applying water to the burning ceiling area near the rear of the house when the building addition

(*continued*)

(*continued*)

roof collapsed, trapping the captain under burning debris. The collapse pushed fire toward the front of the house, which also ignited trapped smoke and sent a fireball rolling toward the front entrance. The fireball engulfed other crews that had entered the house, causing burn injuries to them. Immediate efforts to rescue the trapped captain proved difficult; he was pronounced dead at the scene.

The subsequent investigation involved several representative agencies. One of the agencies was the National Institute of Occupational Safety and Health (NISOH). In their report, several recommendations were made to help minimize the risk of a reoccurrence:

- Ensure that the incident commander (IC) continuously evaluates risk versus gain when determining whether the fire suppression operation will be offensive or defensive.
- Train firefighters to communicate interior conditions to the IC as soon as possible and then provide regular updates.
- Use thermal imaging cameras (TICs) during the initial size-up and search phases of a fire.
- Ensure firefighters open ceilings and overhead concealed spaces as hoselines advance.
- Ensure that team continuity is maintained during fire suppression operations.
- Consider using exit locators such as high-intensity floodlights or flashing strobe lights to guide lost or disoriented firefighters to the exit.
- Train firefighters on the actions to take while waiting to be rescued if they become trapped or disoriented inside a burning structure.
- Consider developing and implementing a system to identify and mark dangerous and/or abandoned structures to improve firefighter safety.

Each of the recommendations should be viewed as preventive, not as a judgment on what did or did not happen in Texas. Given that consideration, each suggestion can be tied to a safety concept discussed in this chapter.

For Discussion:

1. For each recommendation, discuss which of the operational safety elements are at play: procedures, equipment, and/or personnel.
2. For recommendations that related to personnel, which factor needs to be addressed: training, health, and/or attitude?
3. Looking at the fire service as a whole (not necessarily this incident), what attitude issues are present that can lead to—or prevent—similar occurrences?
4. With the limited information available in this case, what specific indicators could have been used to help make risk versus gain judgments?

Note:

This case study was excerpted from NIOSH report F2005-09. The complete NIOSH report is available at www.cdc.gov/NIOSH/fire.

REVIEW QUESTIONS

1. List the three elements that affect workplace safety.
2. Explain the difference between the formal and informal processes.
3. Describe four qualities of a well-written procedure.
4. List and describe the external influences that can affect safety equipment design and purchase.
5. List and briefly describe the three factors that influence a person's ability to act safely.
6. Define risk management.
7. List and explain the five steps of classic risk management.

ADDITIONAL RESOURCES

Angle, James S. *Occupational Safety and Health in the Emergency Services.* Clifton Park, NY: Delmar, a division of Thomson Learning, 1999.

Federal Emergency Management Agency. *Risk Management Practices in the Fire Service,* FA-166. Emmitsburg, MD: United States Fire Administration, December 1996.

Kipp, J. D., and M. E. Loflin. *Emergency Risk Management.* New York: Van Nostrand Reinhold, 1996.

NOTES

1. Rich Duffy and Chuck Soros, *The Safety Officer's Role* (Ashland, MA: Fire Department Safety Officers Association, 1994).
2. Final Report: Southwest Supermarket Fire, Incident #01-045301, Phoenix Fire Department. Available at: http://www.ci.phoenix.az.us/fire/report.pdf. Accessed March 12, 2002.
3. Federal Emergency Management Agency, *Risk Management Practices in the Fire Service,* FA-166 (Emmitsburg, MD: United States Fire Administration, December 1996).

Chapter 3

GUIDING REGULATIONS, CODES, LAWS, STANDARDS, AND PROCEDURES

Learning Objectives

Upon completion of this chapter, you should be able to:

- Explain the motivation for the development of guiding publications.
- List the significant players and their roles in developing guiding publications.
- Define the differences between regulations, codes, laws, and guides.
- List significant publications that can impact the incident safety officer.

INTRODUCTION

Recent events have underscored the need for incident safety officers to understand the many regulations, standards, and procedures that address incident operations and requirements.[1] (For the remainder of this chapter, we use the phrase "guiding publications" to refer to regulations, codes, laws, standards, and procedures.) Incident safety officers have been, and will most likely continue to be, named in criminal and civil courts. When they are, all of the guiding publications become tools to help the ISO avoid litigation—and more importantly—help keep firefighters safe. The majority of the publications were written as a result of a tragic event; therefore, the ISO can use them as a basis to prevent similar tragedies from occurring. It is important to understand that fire officers have participated in the development of guiding publications because they felt it was likely that a similar tragic event may occur in the future—and that probability was unacceptable. The ISO who understands this basic premise can more effectively prevent a similar occurrence.

The creation of guiding publications is better understood when the ISO knows who issued the publication and where it came from. In other words, who are the "players"? Knowing OSHA's role versus NIOSH's is an example. This chapter looks at the roles of the players and the differences in the official publications. We also look at some of the more applicable publications from an ISO perspective.

Note: The information presented here is applicable to situations in the United States. Our firefighter friends in Canada, Mexico, and across the ocean may not find it as applicable, although similar systems probably exist with different names and authorities.

■ **Note**
The majority of the publications were written as a result of a tragic event; therefore, the ISO can use them as a basis to prevent similar tragedies from occurring.

THE PLAYERS

Knowing who the players are can help the ISO understand the breadth of publications and their impact at incidents to help prevent injury and death. Simply stated, hundreds—if not thousands—of established groups are involved with creating guiding publications. In many cases, fire service personnel participate in these groups to help make the document usable in the street world. Let's look at some of the players that have a direct effect on incident activity **(Figure 3-1)**.

National Fire Protection Association (NFPA)

The NFPA was established in 1896 to address a multitude of fire prevention and fire protection issues. The NFPA, while a for-profit association, is

AGENCY	ROLE
NFPA	Development of national minimum consensus standards, codes, and guides. Also collects data and reports trends on a wide range of fire related topics.
OSHA	Develop and enforce the code of federal regulations (CFRs) dealing with occupational safety and health.
NIOSH	Research, investigate, and recommend safe procedures, processes, and habits.
DHS	Develop and implement a national response plan (NRP).
EPA	Issue and enforce regulations and provide training for issues regarding hazardous materials and processes.

Figure 3-1
Significant fire service "players."

recognized for developing consensus standards, guides, and codes for a whole realm of fire-related topics (see "NFPA Standards, Guides, and Codes"). These standards, guides, and codes are developed through committees who are appointed based on the needs of the fire service, private interests, and other technical specialists. Over time, the NFPA has become a data collection resource for many fire-related issues, such as firefighter injury and death statistics, information on civilian fire deaths, and national fire and rescue incident trends. NFPA also offers educational materials, training services, and investigative assistance. It is important to note that NFPA standards are often used to help define what is "acceptable" for fire service equipment, procedures, and professional qualifications. Additionally, NFPA standards could be—and have been—viewed by the courts as *common practice* or *standard of care* when considering legal questions.

■ **Note**

It is important to note that NFPA standards are often used to help define what is "acceptable" for fire service equipment, procedures, and professional qualifications.

NFPA Standards, Guides, and Codes

- *Standards.* A developed body of work that gives minimum consensus direction for procedures, programs, equipment performance, and professional training and qualifications. Standards are written using *mandatory* language.

- *Guides.* A group of publications that NFPA calls *Recommended Practices,* which are written in a language that offers suggestions and in some cases options. Historically, many recommended practices go on to become standards.

- *Codes.* A complete work designed to be adopted as law by an authority having jurisdiction to do so. NFPA's *Life Safety Code* is the best-known.

Code of Federal Regulations (CFRs)
the body of laws enacted by OSHA that are used to help achieve workplace safety

Occupational Safety and Health Administration (OSHA)

OSHA is part of the United States Department of Labor and is tasked with the creation and enforcement of workplace law. OSHA uses the **Code of Federal Regulations (CFRs)** as the body of laws to improve workplace safety. Not all

CFRs are enforceable for the public sector. Individual states adopt OSHA-approved *state plans*. Fire officers should contact their state department of labor to ascertain whether their public entity is covered by a state plan. Regardless, OSHA carries a pretty big stick when it comes to workplace safety and provides a great resource in addressing workplace safety issues.

National Institute for Occupational Safety and Health (NIOSH)

NIOSH is the safety/health research and educational arm of the federal government. NIOSH is actually part of the Centers for Disease Control (CDC) under the Department of Health and Human Services. In 1998 then President Clinton directed—and congress funded—NIOSH to investigate all duty-related firefighter fatalities. NIOSH uses firefighter fatality investigations to help others prevent similar occurrences. These investigative reports can be found at www.cdc.gov/niosh/fire. NIOSH has no enforcement responsibilities, but it can recommend the adjustment or creation of CFRs to OSHA. NIOSH writes several guides for specific hazards to help firefighters better prevent injuries and fatalities.

> **Safety**
> NIOSH uses firefighter fatality investigations to help others prevent similar occurrences. These investigative reports can be found at www.cdc.gov/niosh/fire.

Department of Homeland Security (DHS)

Following the September 11, 2001 attack on the World Trade Center and the Pentagon, President George W. Bush authorized the creation of the Department of Homeland Security (DHS) to better prepare, defend, and respond to terrorist acts within the United States. This cabinet-level department was also tasked with the development of a National Response Plan (NRP) to help manage catastrophic events that are beyond the capabilities of state and local agencies. On February 28, 2003, President Bush issued the Homeland Security Presidential Directive (HSPD-5) that directed the DHS to develop and administer the National Incident Management System (NIMS). Federal grant money to fire departments is tied to their compliance with NIMS. Currently, the Federal Emergency Management Administration (FEMA) and the United States Fire Administration (USFA) fall under the DHS. Given the problematic experiences of dealing with the aftermaths of Hurricanes Katrina, Rita, and Wilma in 2005, the relationship of DHS, FEMA, and the USFA may change.

Environmental Protection Agency (EPA)

The devastating results of hazardous materials release (beginning generally in the 1970s) spurred the creation of EPA to better prevent, respond, and recover from hazmat incidents. EPA has issued many regulations and offers support for hazmat training. It also helps manage superfund monies for cleanup and hazmat training. Even though some state fire agencies are not compelled to follow OSHA CFRs, they are required to follow EPA regulations.

National Institute of Standards and Technology (NIST)

Founded in 1901, NIST is a nonregulatory federal agency within the U.S. Department of Commerce. The mission of NIST is to promote innovation and industrial competitiveness by advancing measurement science, standards, and technology in ways that enhance security and improvement in our quality of life. How does this make NIST a player in the fire service? One arm of NIST, the Building and Fire Research Laboratory (BFRL), is a tremendous resource center that collects a vast amount of information on fire- and building-related subjects. The reports generated by NIST serve as an excellent source of training and education tools to help ISOs understand fire behavior in buildings. Recent reports on the Cook County (Illinois) Office Building fire, the Station House (Rhode Island) fire, and firefighter fatality incidents in Washington, D.C., and Iowa include outstanding visual and written data that are must-reads.[2]

While far from inclusive, the preceding players are responsible for most of the guiding publications we talk about next. Remember that many fire-service–oriented groups also figure in to the "player' group (see "Other Players in the Creation of Guiding Publications").

Other Players in the Creation of Guiding Publications

It is important to note that other players, such as the International Association of Fire Firefighters (IAFF), International Association of Fire Chiefs (IAFC), National Volunteer Fire Council (NVFC), and the Fire Department Safety Officers Association (FDSOA)—to name a few—are intimately involved in the processes to create, alter, and implement guiding publications.

DEFINING THE GUIDING PUBLICATIONS

There are thousands of publications that may have an impact on fire and emergency service personnel. It is important to understand not only the differences among these publications and their applicability, but also that all of them have a common goal: safety. At the street level, fire service personnel often throw around terms such as "codes" and "standards" interchangeably. Doing so may cause confusion and certainly misrepresents the specific applicability of each publication. Let's look at some of the intricacies of these publications.[3]

Regulations

Regulations typically outline details and procedures that have the force of law issued by an executive government authority. OSHA CFRs and EPA regulations are examples.

Codes

A code is a work of law established or adopted by a rule-making authority. Codes serve to regulate an approach, system, or topic for which it is written. The *Uniform Fire Code* and the *Life Safety Code* are examples.

Standards

■ **Note**

To have the effect of law, a standard must be adopted by an authority with the legal responsibility to enact the standard as law (that is, promulgate it).

NFPA standards are perhaps the most familiar to fire service personnel. The term "standard" can apply to any set of rules, procedures, or professional measurements that are established by an authority. In the case of NFPA standards, a formal consensus approach is used. To have the effect of law, a standard must be adopted by an authority with the legal responsibility to enact the standard as law (that is, promulgate it). For example, a city may enact a local ordinance that adopts a standard.

Laws

statutory law
laws promulgated to deal with civil and criminal matters

case law
law that refers to a precedent (a ruling by a judge or a specific court proceeding) established over time through the judicial process

Laws are enforceable rules of conduct that help protect a society. From a legal prospective, laws are divided into *statutory law* and *case law*. A **statutory law** deals with rules of conduct in civil and criminal matters. **Case law** refers to a precedent established over time through the judicial process. As a (fictitious) example, an incident commander is charged with criminally negligent homicide (statutory law) following a firefighter fatality. In the court proceedings, the IC may be found not guilty because of a precedent set by *United States v. Gaubert*, 486 U.S. 315, 111 S.Ct. 1267, 1991 (case law).

While it may sound confusing, some jurisdictions use the term "code" for their statutory laws. Local research can help fire officers understand how their jurisdictions use the two terms.

Guides

■ **Note**

Guides do not have the impact of a law, but they can be used in negligent cases to provide evidence of general duty or standard of care.

Publications that offer procedures, directions, or standard of care as a reasonable means to address a condition or situation are called *guides*. Guides do not have the impact of a law, but they can be used in negligent cases to provide evidence of general duty or standard of care. NIOSH has written several guides to address firefighter safety issues, and occasionally it issues an *alert,* which is another form of a guide. Alerts are issued in response to a disturbing trend of injuries or deaths by a specific cause (like the failure of truss systems during structural fires) and typically outline case studies, technical research, and preventive measures. Textbooks (like this one) can also be considered guides and are often cited in investigations and legal proceedings to establish a standard of care.

PUBLICATIONS THAT AFFECT THE ISO

Once you understand the roles of the players and guiding publications in emergency services, you can look at individual publications that may have a direct impact on firefighter safety. To cite them all here would take volumes; it is a compelling list. Instead, presented here are some of the more important documents that apply to incident handling and firefighter safety. The effective ISO will spend time researching the depth of these publications—beyond what is offered here.

NFPA 1500, Standard on Fire Department Occupational Safety and Health Program

First published in 1987, the NFPA 1500 standard has become the "mother ship" of fire department safety and health because it ties together many other NFPA standards by referring to them by their applications. NFPA standards for professional qualifications, protective equipment, tools, apparatus, incident management, and training are all cited in the document. NFPA 1500 continues to evolve through the efforts of the Technical Committee made up of a diverse group of stakeholders. The current edition is divided into twelve chapters and an extensive annex section with explanatory information, checklists, and examples **(Figure 3-2)**.

Of particular note to the ISO, NFPA 1500 addresses safety at the incident scene in Chapter 8, Emergency Operations. The chapter is broken into 11 areas:

1. Incident Management
2. Communications
3. Risk Management During Emergency Operations
4. Personnel Accountability During Emergency Operations
5. Members Operating at Emergency Incidents
6. Control Zones
7. Roadway Incidents
8. Rapid Intervention for Rescue of Members
9. Rehabilitation During Emergency Operations
10. Scenes of Violence, Civil Unrest, or Terrorism
11. Post-Incident Analysis

The ISO should be very familiar with NFPA 1500—and stay informed of changes as the document evolves.

NFPA 1521, Fire Department Safety Officer

Perhaps the best starting place for understanding the roles, responsibilities, functions, and authorities of an incident safety officer is NFPA 1521. Although

NFPA 1500 CHAPTER ORGANIZATION (2007 EDITION)

CHAPTER 1 :   ADMINISTRATION
CHAPTER 2 :   REFERENCED PUBLICATIONS
CHAPTER 3 :   DEFINITIONS
CHAPTER 4 :   FIRE DEPARTMENT ADMINISTRATION
CHAPTER 5 :   TRAINING, EDUCATION, AND PROFESSIONAL DEVELOPMENT
CHAPTER 6 :   FIRE APPARATUS, EQUIPMENT, AND DRIVER/OPERATORS
CHAPTER 7 :   PROTECTIVE CLOTHING AND PROTECTIVE EQUIPMENT
CHAPTER 8 :   EMERGENCY OPERATIONS
CHAPTER 9 :   FACILITY SAFETY
CHAPTER 10 :   MEDICAL AND PHYSICAL REQUIREMENTS
CHAPTER 11 :   MEMBER ASSISTANCE AND WELLNESS PROGRAMS
CHAPTER 12 :   CRITICAL INCIDENT STRESS PROGRAM
ANNEX A :   EXPLANATORY MATERIAL
ANNEX B :   FIRE SERVICE OCCUPATIONAL SAFETY AND HEALTH PROGRAM WORKSHEET
ANNEX C :   BUILDING HAZARD ASSESSMENT
ANNEX D :   RISK MANAGEMENT PLAN FACTORS
ANNEX E :   FIREFIGHTER SAFETY AT WILDLAND FIRES
ANNEX F :   HAZARDOUS MATERIALS PPE INFORMATION
ANNEX G :   SAMPLE FACILITY INSPECTOR CHECKLISTS
ANNEX H :   REFERENCED PUBLICATIONS

Figure 3-2   *NFPA 1500, Fire Department Occupational Safety and Health Program, 2007 Edition.*

the title "safety officer" is generic, NFPA 1521 adds specificity to it by defining various types of safety officers. The types of safety officers, as defined by NFPA, are:

- *Health and safety officer.* The member of fire department assigned and authorized by the fire chief as the manager of the occupational safety and health program.

- *Incident safety officer.* A member of the command staff responsible for monitoring and assessing safety hazards or unsafe situations and for developing measures to ensure personnel safety at the scene of an incident.

- *Assistant incident safety officer (ASO).* An individual appointed to respond or assigned at an incident scene by the incident commander to assist the incident safety officer in the performance of the ISO functions.

The standard goes on to define the assignment, authorities, and qualifications of each role. It recognizes that one person may be the HSO and another the ISO and that the authorities of the ISO are equally applied to ASOs.

NFPA Standard 1521 also cites the assignment, authorities, and qualifications of an ISO, although some highlights are emphasized:

- The fire department shall develop an ISO response system that ensures that a predesignated ISO responds or is appointed to all working incidents.
- The ISO shall utilize ASOs when the size, scope, or technical merit of an incident warrants doing so.
- The ISO shall meet the requirements of a Fire Officer I as specified in NFPA 1021, *Standard for Fire Officer Professional Qualifications*.
- The ISO shall have a working knowledge of firefighting safety and health hazards, building construction, fire behavior, rehabilitation strategies, personnel accountability systems, and abilities to perform the functions of the ISO as described in Chapter 6 of the standard (Functions of the ISO).
- The ISO shall have the authority of the incident commander to stop, alter, or suspend activities that present an imminent threat to firefighters.
- The ISO shall inform the IC when he or she has stopped, altered, or suspended an activity.

The standard also addresses specific functions of the ISO—duties and responsibilities that the ISO should be addressing. The ISO function chapter is divided into the following seven areas:

1. Incident management system (IMS) (addresses how the ISO fits into the IMS and requires that the ISO is easily identifiable by a unique item, such as a vest or helmet)
2. General incident safety (a list of safety items that need to be addressed regardless of the type of incident, such as risk/benefit thinking, rehab, personnel accountability, zoning, action plan review, communications, and traffic)
3. Fire suppression
4. Emergency medical service operations
5. Hazardous materials (requires that a hazmat incident requires a hazmat technician as the ISO or the utilization of an ASO who is a technician level)
6. Special operations
7. Postincident responsibilities

As with NFPA 1500, the 1521 standard continues to evolve. Staying aware of the requirements of 1521 is essential for the ISO to remain effective. Since NFPA 1521 includes some professional qualification requirements, there has been talk of creating a Pro-Qual (professional qualification) standard for safety officers (tentatively called "safety technician") to go along with the other one thousand series of NFPA documents.

IDLH
an acronym given to environments that are immediately dangerous to life and health

■ **Note**
The series known as Title 29 CFR includes numerous subtitles that are specific to public sector members who engage in rescues and exposure to environments that are *immediately dangerous to life and health* (IDLH).

OSHA TITLE 29 CFR

OSHA's primary focus area is in the private sector. Some CFRs, however, are intended to apply to the public sector as well. In some cases, the CFR specifically speaks to the rescue of employees engaged in certain activities (like confined space work). The series known as Title 29 CFR includes numerous subtitles that are specific to public sector members who engage in rescues and exposure to environments that are *immediately dangerous to life and health* (**IDLH**). The following CFRs may have some impact on the functions of the ISO:

- 29 CFR 1910.120, *Hazardous Waste Operations & Emergency Response Solutions*
- 29 CFR 1910.134, *Respiratory Protection*
- 29 CFR 1910.146, *Permit-Required Confined Spaces*
- 29 CFR 1910.147, *The Control of Hazardous Energy (lockout/tag out)*
- 29 CFR 1910.1030, *Blood-Borne Pathogens*
- 29 CFR 1910.1200, *Hazard Communication*
- 29 CFR 1910.1926, *Excavations, Trenching Operations*

As they relate to the ISO, the OSHA CFRs emphasize the need to have a site safety plan for operations involving hazmat, confined spaces, trenches, and hazardous energy. When the fire department is engaged in rescue activities involving these, the ISO should develop a site safety plan (in writing) and present a briefing to those working the incident. The development of safety briefings is discussed later.

NIOSH Publication 2004-144, Protecting Emergency Responders, Volume 3

The subtitle of this publication is *"Safety Management in Disaster and Terrorism Response."* Eight chapters in this reference can help the ISO better understand safety issues in all types of disasterlike incidents. The September 11 World Trade Center attack was used as a "lessons-learned" event to make this publication more usable to the initial incident responder. Chapter 7 of the publication addresses an integrated, incident-wide safety management approach that forms a template for ISOs. The publication is available at www.cdc.niosh/docs/2004-144.

NIOSH Alert, Preventing Injuries and Deaths of Firefighters due to Truss System Failures

This alert was issued in April of 2005 in response to 25 firefighter fatalities in five years due to truss roof collapse. The alert cites several case studies and includes recommendations from several noted fire service leaders on proactive strategies and tactics for fires in truss buildings. The 25 pages should be a must-read for all ISO—and fire officers! It is also available at the NIOSH Web site.

NIST Special Publication SP-1021, Cook County Administration Building Fire, October 17, 2003

The National Institute of Standards and Technology has released a comprehensive video DVD that summarized numerous heat release experiments to help understand the fire dynamics associated with the Cook County High-rise fire in Chicago. The information in this DVD should be mandatory training for all ISOs. Of particular note, the DVD shows actual and simulated fire dynamics that can be expected from office work-station fires. ISOs can glean useful information on rapid fire and smoke spread from this outstanding publication.

SUMMARY

Recent events have underscored the need for incident safety officers to understand the many regulations, standards, and procedures that address incident operations and requirements. The majority of these guiding publications have been developed following a catastrophic event or to address disturbing injury and death trends to prevent future occurrences. Several key players or agencies have taken responsibility to develop and implement various guiding publications. They include the NFPA, OSHA, NIOSH, DHS, and the EPA. Of note, NFPA standards are often used to define what is acceptable in the way of equipment, procedures, and professional requirements for the fire service. Regulations, codes, standards, laws, and guides are terms that are mistakenly used interchangeably; the ISO must understand the definition and intent of each. There are numerous publications that can impact the ISO. Understanding what these publications are and the intent of each can help ISOs better perform their functions—and help prevent injuries and deaths.

KEY TERMS

Case law   Law that refers to a precedent (a ruling by a judge or a specific court proceeding) established over time through the judicial process.

Code of Federal Regulations (CFRs)   The body of laws enacted by OSHA that are used to help achieve workplace safety.

IDLH   An acronym given to environments that are immediately dangerous to life and health.

Statutory law   Laws promulgated to deal with civil and criminal matters.

POINTS TO PONDER

A New York Tragedy

On September 25, 2001, Firefighter Trainee Bradley Golden died during a live fire training exercise in Lairdsville, New York. Assistant Chief Alan Baird III, the designated instructor for the training exercise, was charged with second-degree manslaughter and later found guilty of criminally negligent homicide by an 11-person jury.

The case against Baird arose out of a live fire rapid intervention exercise during which Golden and another firefighter posed as victims. Golden, who had only been a volunteer firefighter for a few weeks, had never had any formal training and had never worn an SCBA in a live fire environment. Despite this, Golden and another firefighter were placed in the second-floor front bedroom of a residential duplex and covered with debris to simulate entrapment. A burn barrel was ignited on the second floor to develop smoke, and shortly after Baird ignited a foam mattress on the first floor. Within minutes of the mattress ignition, flames rolled across the ceiling, up the stairway, and out the windows of the front bedroom where the simulated victims were. No hose lines had been stretched prior to the ignition of the fires. A total of three firefighters were trapped on the second floor with limited egress options. One firefighter self-extricated by prying open a boarded-up window and the others were removed from the building by crews that had been staged for the drill. Golden was found unresponsive and was transported to the hospital, where he was pronounced dead. The other two firefighters suffered burns and were airlifted to the hospital.

In his defense, Baird testified that he was not the officer in charge and that the incident commander and designated safety officer were aware of the drill objectives and knew what fires were to be set. Baird pointed out that he was not the highest-ranking officer on scene. Baird also testified that at the time of the exercise he was not aware of the NFPA standards that address live fire training. Additionally, Baird testified that the training was being accomplished following procedures that they used at a previous live fire training event several years earlier that was under the direction of a state fire inspector. Baird's counsel argued that Baird should not be held accountable for not knowing about the NFPA standards and that he should have never been charged because the tragedy was an "accident." Prosecutors cited NFPA 1403, *Standard for Live Fire Training* in their arguments and ultimately convinced the jury that Baird violated many nationally known standards.

For Discussion:

1. What role does NFPA standards play in the defense and prosecution of fire officers in criminal matters or in the judgment of civil cases?

2. Why do you think the prosecution focused on the designated instructor?

3. What accountability does the incident commander and safety officer have in similar situations?

4. Some feel that Golden's death was an unfortunate accident and that this type of dangerous training prepares firefighters for the real risks they face at fires. What arguments can you prepare to defend or counter this feeling?

Note:

This case study was developed using information from the NIOSH fatality investigation report and assistance from Attorney David C. Comstock, Jr., J.D., Chief Fire Officer, Western Reserve Joint Fire District, Poland, Ohio.

REVIEW QUESTIONS

1. What has typically motivated the establishment of guiding publications?

2. How are OSHA and NIOSH different?

3. What is the significance of the U.S. Department of Homeland Security to the fire service?

4. Define regulations, codes, laws, and guides.

5. List the 11 topical areas in the NFPA 1500 chapter on emergency operations.

6. What are the seven topical areas listed in NFPA 1521 for the functions of an ISO?

7. What does "IDLH" stand for?

8. What responsibility does the ISO have in the use of OSHA Title 29 CFRs?

ADDITIONAL RESOURCES

Callahan, Timothy, *Fire Service and the Law,* 2nd ed. (Quincy, MA: National Fire Protection Association, 1987).

National Institute for Occupational Safety and Health (NIOSH) Web site: Available at: http://www.cdc.gov/niosh.

National Institute of Standards and Technology (NIST) Web site: Available at: http://www.nist.gov.

NFPA 1500, *Standard on Fire Department Occupational Safety and Health Program* (Quincy, MA: National Fire Protection Association, 2007.

NFPA 1521, *Standard on Fire Department Safety Officer* (Quincy, MA: National Fire Protection Association, 2008 [proposed]).

Occupational Safety and Health Administration (OSHA) Web site: Available at: http://www.osha.gov.

NOTES

1. Author's observation and study of numerous court cases, OSHA violations, and fatality investigation reports.

2. NIST reports can be found at www.cdc.gov/niosh/fire.

3. Author's Note: A special thanks to attorney and Fire Chief David Comstock, Jr., Western Reserve Joint Fire District, Poland, Ohio, for his assistance in the development and review of this material.

Chapter 4

DESIGNING AN INCIDENT SAFETY OFFICER SYSTEM

Learning Objectives

Upon completion of this chapter, you should be able to:

- Discuss the reasoning for preplanning the response of an incident safety officer.
- List four examples of when an automatic ISO response should take place.
- List four examples of when an incident commander should automatically delegate the safety responsibility to an ISO.
- List and discuss the advantages and disadvantages of using various methods to ensure that an ISO arrives on scene.
- Discuss the authorities suggested for incident safety officers by NFPA standards.
- List several tools that will help the ISO be effective on scene.

ISO DESIGN OVERVIEW

The design and implementation of a fire department incident safety officer program or system can make the difference in whether or not a program will be effective. When the first edition of this book came out, many fire departments took suggestions from it to help make their ISO system work. Since then, departments have discovered through trial and error ways to maximize the effectiveness of their ISO program. The options now available for an ISO system design are numerous—and endless—based on fire department size, deployment strategy, and unique needs. Nevertheless, the design of the ISO program should address some key questions:

- Who responds and fills the ISO role?
- What type of incidents necessitate the use of an ISO?
- What tools and training are necessary to maximize ISO effectiveness?

NFPA

Standard operating procedures shall define criteria for the response of a pre-designated incident safety officer.

When designing a system, it is important to keep in mind the suggested requirements from NFPA 1521. The standard states that "Standard operating procedures shall define criteria for the response of a pre-designated incident safety officer."[1] The standard further explains this requirement in an appendix note:

A fire department should develop response procedures that ensues that a pre-designated ISO, independent of the IC, responds automatically to pre-designated incidents.

In this chapter, we explore the rationale for the NFPA requirement as well as suggestions on how to design an ISO system.

PROACTIVE ISO RESPONSE

■ **Note**
If an incident commander is truly going to make a difference at an incident scene, the delegation of the safety function needs to be *proactive*.

Most incident commanders, fire chiefs, and firefighters would agree that incident safety officers are necessary and valuable at significant or complex incidents. However, the discovery that an incident is significant or complex typically comes after incident operations are underway and the incident management system has been set up. This leads to a situation in which the delegation of an ISO is *reactive*. For example, an emergency call is received and a programmed response is initiated. Once on scene, the incident command system is implemented and actions are taken to begin mitigation. If the incident is significant or overcomes the initial response, additional resources are requested. At about this time, an incident safety officer (ISO) becomes a consideration for the incident commander (IC). In this case, the IC is reactive in the delegation of firefighter safety duties. If an incident commander is truly going to make a difference at an incident scene, the delegation of the safety function needs to be *proactive*. To be proactive in the delegation and placement of an ISO, the fire department needs to *preplan* the ISO response. Let's explore *why* you need to preplan an ISO response and cover the "when, where, and how" of an effective ISO system.

PREPLANNING THE ISO RESPONSE

A few incident commanders believe that any fire officer should be able to fill the ISO position, at anytime, under any circumstance, at the will and want of the incident commander—therefore, the agency really doesn't need to create an ISO system. Not only is this thinking flawed, it is dangerous. Just as incident commanders have various levels of knowledge and expertise, so do other fire officers. Likewise, the requirements necessary to be a fire officer may change from department to department—a problem if the need for mutual aid arises. Further, the emphasis placed on safety may vary from one IC to another. As stated in Chapter 1, firefighter death and injury statistics suggest that more can be done to improve firefighter safety. Those same statistics show that the majority of deaths and injuries on the fireground occur at residential structure fires.[2] Logic (as opposed to traditional fireground thinking) suggests that we send an ISO to all residential fires. We know, however, that some incident commanders are thinking, "Residential fire . . . couple engines . . . a dozen people . . . we can handle that . . . we don't need a safety officer." The statistics show otherwise. Consider:

! Safety
The ISO is most effective when he or she arrives early in the incident.

The ISO is most effective when he or she arrives early in the incident.

NIOSH, which has a federal mandate to investigate all firefighter fireground fatalities, has recognized the need to have an ISO appointed early. A reoccurring recommendation has been cited in these firefighter investigations:

Ensure that a separate ISO, independent from the IC, is appointed early at an incident or responds automatically to pre-designated fires.[3]

A few pragmatic graphs (as opposed to scientific ones), based on a typical residential working fire, can also illustrate the need to have an ISO available early at fires.

Graph 1: Environmental Change

As applied to a residential structure fire, "environmental change" means fire propagation, building degradation, and smoke volatility. As you can see in **Figure 4-1**, the rate of change is measured and rated over time. During a fire in a structure, there is actually a routine, or even a rapid, rate of change upon arrival of fire crews.

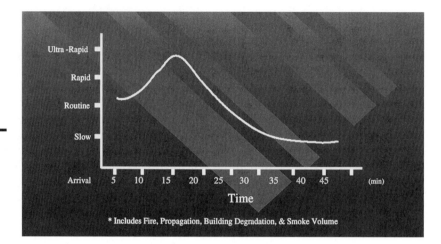

Ultra -Rapid
Rapid
Routine
Slow
Arrival　5　10　15　20　25　30　35　40　45　(min)
Time
* Includes Fire, Propagation, Building Degradation, & Smoke Volume

Figure 4-1 *The rapidly changing environment early in a fire is grounds for the early appointment of a safety officer.*

Granted, this depends on response time. In most cases, the fire is intensifying upon arrival (that's probably why someone called 9-1-1). At arrival (zero on the time line), the fire is starting to develop significantly. With that comes smoke production as more contents (fuel) become heated. Equally, the building itself is being attacked and thus becoming structurally degraded—especially in lightweight structures. If flashover happens prior to fire department control efforts, the environmental change becomes ultrarapid. Other events can cause ultrarapid change. Smoke explosions, backdrafts, partial collapse, the presence of accelerant fuels, and other phenomena can all cause ultrarapid change in the firefighting environment. Once control efforts have begun, this rate of change can be stabilized and even reduced. It would seem to make sense to have an ISO on hand early to help evaluate hazards during these periods of rapid and ultrarapid change.

Graph 2: Fireground Activity

Another area that makes an early ISO assignment advantageous is the amount of fireground activity. Upon arrival at a typical residential fire, you can argue that no tasks are being performed (**Figure 4-2**). Even though all responders are going through their personal size-ups and potential incident commanders are performing a lot of mental activity, no crews are performing physical, active tasks in the dangerous environment upon arrival. As on-scene time passes, the activities needed to control the incident increase: Perhaps one to three tasks are being accomplished based on the number of resources that arrive initially. Within 20 minutes, the IC may be orchestrating seven to ten simultaneous assignments, which may include search and rescue, exposure protection, ventilation, preparing attack and backup lines, rapid intervention

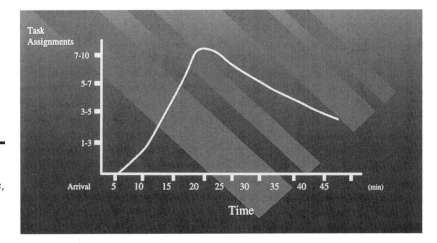

Figure 4-2   *The higher the number of task assignments, the greater is the need for a safety officer.*

planning, etc. For some departments, the simple task of water supply is a significant chore that incorporates two or three task assignments that are going on simultaneously. The point is simple: About 15 to 20 minutes into the incident, a lot of tasks are going on all at once. A key to preventing injuries is enabling the incident commander to monitor the numerous simultaneous activities during the first 20 minutes after arrival. The good incident commander wants an ISO appointed early on in the incident. Wouldn't it be great if you, as the IC, knew that and an ISO was predesignated and responding before five or six tasks were being performed?

> **! Safety**
> The good incident commander wants an ISO appointed early on in the incident.

Graph 3: Relative Danger to Firefighters

Practically speaking, the first-arriving firefighters at a working fire should perform some kind of risk/benefit analysis (see Chapter 2 for risk management concepts). This initial risk analysis comes with a certain amount of risk—simply because the firefighters may not know the full extent of the dangers at hand upon arrival. Firefighters actually begin taking risks upon arrival—or at "zero time." That is, at times, risks are taken just to determine the risk of the incident! Here is an actual tragic example: The Denver Fire Department was called to assist the police department on a welfare check. A firefighter was assigned to climb a ladder to check for a possibly unlocked, upstairs window. He reached the top of the ladder and, without warning, was shot by the occupant of the house!

Early on in a fire incident, risk taking can become extreme or high (**Figure 4-3**). The perfect example of extreme risk taking is search and rescue. In a typical scenario, a first-arriving crew starts setting up for an aggressive interior attack; they're laying supply lines in, pulling attack lines, getting air

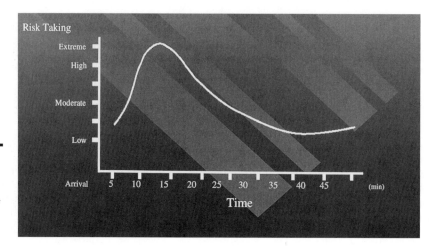

Figure 4-3
Firefighter risk taking is usually greatest in the early stages of an incident.

packs on—preparing for an offensive attack. All of a sudden the first-due officer comes from the back of the building after doing a "360" and reports that there is an immediate need for rescue. Three victims are hanging on the third floor balcony and they need to be evacuated "right now." The crew drops the planned attack and begins taking risks to get to the people. It is early on in an incident, when risk taking is high. Volunteer, paid on-call, or pager-notified fire departments, who allow firefighters to respond directly to the incident scene, may face an even greater degree of risk taking by reason of having firefighters on-scene prior to apparatus and equipment. Once again, the point is simple: Risks are usually greater early on in an incident; therefore, that is when a safety officer is needed.

Overlapping the Graphs

Fire and rescue departments that wish to make a difference in the reduction of firefighter injuries and death develop a system to get the ISO on scene or appointed early on in an incident.

The overlapped graphs in **Figure 4-4** clearly show that the first 20 minutes of an incident warrant close monitoring of firefighters and firefighting operations. The early appointment of an ISO gives the incident commander another set of eyes, another viewpoint, and another consultant. The logic illustrated by the graphs is sound. Although no accurate data suggest that firefighter deaths and injuries happen early at an incident, most fire officers would agree that the most dangerous time at an incident is the critical point when they are making aggressive efforts to rescue victims and stop a rapidly growing fire. The only way we are going to get an ISO to an incident early is to predefine *when* an ISO is required to be appointed and design a system to make sure a trained ISO arrives on-scene early.

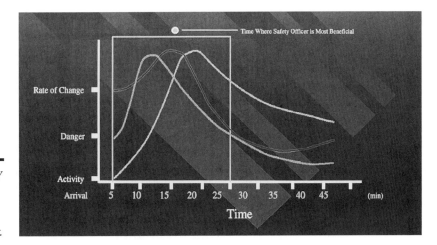

Figure 4-4   *The early assignment of an incident safety officer is essential for firefighter safety.*

When Does the ISO Respond?

If, as suggested, an incident safety officer should be present on all working residential fires, common sense suggests that an ISO should also be planned for highly technical or complex incidents. For all these reasons, the response of an ISO should be preplanned (proactive). It is not suggested that the incident commander's ability and authority to make decisions be taken away. What is suggested is that firefighter injury statistics show that we need to have a dedicated ISO more often and sooner. If an IC can get an extra set of eyes on-scene sooner, firefighters are likely to be more able to manage hazards better and therefore more safely. This helps to reduce death and injury occurrence. In defining *when* an ISO is required to respond, the individual fire and rescue department should develop guidelines for an *automatic ISO response* as well as guidelines, or circumstances, that mandate an *automatic ISO delegation* during the incident.

> **Safety**
> firefighter injury statistics show that we need to have a dedicated ISO more often and sooner.

Automatic FDISO Response

The goal is to have a predesignated, trained ISO programmed to respond to certain incidents with a high risk to firefighters (again, refer to the risk management concepts in Chapter 2). Generally speaking, the types of incidents that may require an automatic ISO response are as follows.

Residential or Commercial Structure Fires   Upon receipt of a 9-1-1 or other report of a fire in a residential or commercial fire, an ISO should be programmed into the response **(Figure 4-5)**. For some departments, dispatch personnel may need to add a level of screening so that a "burnt toast" call is not classified as a structure fire. Some may argue that an automatic response would lead to

Figure 4-5 *The report of an actual hostile fire should trigger an ISO response.*

having an ISO at every fire, something that a large department (cost) or small department (available staffing) may not find acceptable. Unfortunately, statistics do not justify this argument. Residential fires hurt and kill firefighters! Once we, as a fire service, force a significant decline in injuries, then maybe sending an ISO to every structure fire is not necessary. Until then, the merits of the requirement outweigh its cost. In fact, small or critically staffed fire agencies can make the argument that, the fewer people who are available, the more important an ISO becomes; in essence, each person on-scene is taking a greater risk than they would in an appropriately staffed response.

Commercial buildings present a profusion of hazards that require immediate understanding—again requiring an early ISO response. High-rise building fires also have unique hazards that need to be addressed early on in an incident.

Wildland-Interface Fires   Fires in the wildland or at the wildland/urban interface (I-Zone) can cause unique problems and present unusual choices for first responders. First-arriving company officers have to evaluate fuels, weather, topography, fire conditions, access, and the defensibility of threatened structures **(Figure 4-6)**. This combination leads to an environment well suited for the partnership of an incident commander and incident safety officer early on in the incident. In some jurisdictions, the time of year and local fuel conditions may indicate the need for an automatic ISO response. For example,

Figure 4-6 *Fires in the wildland, or I-Zone, should trigger an automatic ISO response.*

when fire danger is rated high or extreme by the forest service or other agency, the local fire department may "bump up" its resource response to include the automatic response of an ISO to any fire report.

Specialty Team Incidents   Many fire and rescue departments have specific-function teams designed to handle the array of incidents outside the scope of standard engine and truck company functions **(Figure 4-7)**. Examples are hazmat, dive rescue, heavy (urban) rescue, rope rescue, and wildland "hot shot" teams. Often, the activation of these teams takes a separate request or page, and an ISO should be included with this activation. Some of these teams include an ISO specially trained to interface with the needs of the particular team. These ISOs could be termed assistant safety officer–hazmat or ASO–tech rescue. The now famous urban search and rescue (USAR) teams around the nation have specially trained safety officers who travel with the team. If the team does not include its own ASO, a basic-trained ISO can function as a "common sense" consultant to the team, support personnel, and incident commander prior to the appointment of the specially trained ASO.

Target Hazard Incidents   In nearly every city, local firefighters can name a building where they hope they never get a fire—" 'cause it'll be a killer." One advantage of computer-aided dispatch (CAD) systems is the ability to flag

Figure 4-7 *The activation of specialty teams should also activate ISO response.*

locations and occupancies that present unique hazards to firefighters. If an incident is reported at one of these locations, an ISO can be preprogrammed into CAD to respond. Examples may include chemical and industrial plants, historical buildings, stadiums, underground structures, difficult- or limited-access occupancies, and the like.

Aircraft Incidents The potential for high loss of life, mass fire, and the release of hazardous materials is prevalent in aircraft incidents. Hence, an ISO should be included in this type of response. Departments with airfield aircraft rescue and firefighting (ARFF) responsibilities have specialized equipment and procedures to minimize the associated risks. An ALERT 3 (crash) can stretch these safety measures, especially if the airport has an ALERT 3 with an out-of-index aircraft ("indexing" of aircraft refers to its size and passenger load). In Fresno, California, for example, a military contract Lear jet crash-landed on a street in the middle of town and instantaneously started an aircraft, apartment, and commercial building fire, as well as a hazmat and collapse incident. At this type of incident, an incident commander would surely embrace the early arrival of a designated incident safety officer.

Weather Extremes Firefighting efforts in subfreezing temperatures or in high-humidity and high-temperature environments present additional hazards to

Figure 4-8 *Incidents during extreme weather should include the response of an ISO.*

firefighters **(Figure 4-8)**. Changes in strategies, tactics, and action plans may be required just because of the weather. If firefighters are unaccustomed to working in a given weather extreme, injury risk rises. As an example, for the Colorado Front Range area (Denver, Colorado Springs), temperatures on either side of 0 and 100 degrees Fahrenheit happen just a few days a year; therefore firefighters may not be acclimated to the extremes. In this case, accelerated rehabilitation and monitoring of firefighters is a solid argument for an automatic response by an ISO. It makes sense to get an ISO on-scene early to help monitor the risks associated with the added stress of excessive snow or heat. On the other hand, firefighters in parts of Minnesota or upstate New York may think it silly to send an ISO just because it is cold and the snow is a little deep, but on a very hot, humid day, they might need someone monitoring their reaction to the temperature. Our firefighter friends in Florida and Arizona are probably well versed in hot weather firefighting, but throw them a 20-degree–Fahrenheit temperature and watch the stress! Acclimation is the key: If the weather becomes extreme in terms of *your* jurisdiction, then consider adding an ISO to the response.

Automatic ISO Delegation

"Plan for the worst" is a personal code that most responding fire officers live by. The odds are that fire officers have enough resources to handle most "routine" situations. Occasionally, however, they are caught short or, worse, think the situation has been handled—only to have "Murphy" show up. In these cases, it pays to have a predefined set of circumstances that lead to the automatic delegation of the ISO role. By having these situations predefined, the incident commander is sure to get the valuable assistance that an ISO can provide for escalating or multiple-risk situations. Let's look at some examples.

Working Incidents   Generally speaking, a *working incident* can be defined as one in which the first-due, on-scene resources are 100 percent committed and more are needed. In these cases, the control effort is behind the power curve, which leads to greater risk taking by personnel **(Figure 4-9)**. The IC's call for additional assistance becomes the indicator that an ISO is needed. Although language varies among jurisdictions, the striking of a second alarm can serve as the impetus for an automatic ISO delegation.

A Span of Control in Excess of Three   Most incident management systems recognize an ideal span of control of five or less for emergency operations. Those well versed in IMS, however, choose to delegate sections or functions before the one-to-five mark. Once the delegation point is reached, the IC should include an ISO. If the IMS is handling an incident using groups and divisions, the addition of the fourth group or division should be the signal to delegate an ISO. Also, the IC's decision to delegate the existing groups or divisions into

Figure 4-9 *Working fires require the rapid delegation of safety responsibilities.*

an operations section could be the signal to appoint an ISO. Regardless of the type of IMS used, the department should include a "mark" that signifies the point at which an ISO is automatically assigned into the system.

Mutual Aid Request   For smaller departments, the appointment of an ISO whenever mutual aid is requested can be a firefighter saver. Even though many mutual-aid companies work well together **(Figure 4-10)**, the mutual-aid environment asks for firefighters with different cultures, tactics, and equipment to work together. If for no other reason, an ISO should be appointed to monitor and ensure that the action plan is being carried out as intended by the visiting team. This makes sense for moral reasons, if not for legal reasons. (When mutual aid systems train and respond together on a regular basis (say, monthly), the need for an automatic ISO delegation to all mutual aid requests is not as great.)

Firefighter Down, Missing, or Injured   A seemingly simple incident can get turned upside down and shaken loose when a firefighter emergency takes place. The well prepared fire and rescue department has a firefighter emergency plan and a rapid intervention crew or company (RIC) in place to deal with such an emergency—although they hope never to use it. One key part of the plan should be the immediate appointment of an ISO to help the IC implement it. An ISO may also be required by the local jurisdiction to start the accident documentation and investigation process.

Incident Commander Discretion   Incident management systems recognize that the IC is ultimately responsible for firefighter safety and the incident outcome. At

Figure 4-10 *Mutual aid incidents give rise to an array of safety concerns.*

Incident Safety Officer Utilization	
Automatic Response	**Automatic Delegation**
• Commercial/residential fires	• "Working" incidents
• Wildland/interface fires	• Growing span of control
• Special-Team incidents	• Mutual-aid incidents
• Target hazards	• Firefighter down/missing/trapped
• Aircraft incidents	
• Extreme weather	

Figure 4-11 *Automatic response and automatic delegation responses summarized.*

anytime and under any circumstance, the IC can delegate the ISO function. For this reason, a trained ISO needs to be available for delegation. In fact, a department health and safety officer or on-duty ISO should have some discretion to respond to an incident. In these cases, an IC may question or take exception to the ISO's showing up, but their mutual concern for safety should override personality differences and/or turf protection. A well written policy can help avoid differences.

A summary of automatic ISO responses and delegated ISO responses is shown in **Figure 4-11**.

WHERE DOES THE ISO COME FROM?

Once a fire and rescue department decides to pre-plan its ISO response, a system to get an ISO on-scene is imperative. Who is this person? Where does he or she come from? How is he or she alerted? There are probably as many answers to these questions as there are fire departments. Let's look at some typical arrangements.

Training or Safety Officer(s) on Call   Of all the systems to get an ISO on to the scene, having the training or safety officer on call seems to be very popular[4]—although it may not be the most effective. In these systems, the officer can either monitor incident activity and self-dispatch or count on a pager to be notified when an ISO is needed.

- *Advantages:* This system is popular because many departments believe that the training officer is a readily available fire officer who knows how the department operates. Further, the responsibilities of training and safety are embodied in one person—an attractive package to sell in the preparation of budgets, staffing reports, and other planning. Also, training officers often have the radios, vehicles, and other tools that make their transition into the ISO role quick. Individual training officers may like this option because it gives them a chance to get out of the daily routine and help maintain their incident skills.

- *Disadvantages:* Training officers are not always available around the clock, putting a burden on the department and the officer. Often, the training officer is actually the Training/Safety/Infection Control/ Research and Development/Special Projects/Recruitment/Wellness/ Quality Control/Accreditation Officer for the department. Wearing all these hats usually comes at the expense of not wearing any one of them well. In these cases, the training officer is probably better suited to train a contingent of ISOs to be available for response.

Health and Safety Committee Members In this system, the members of the department's health and safety committee are trained to serve as on-duty ISOs. Typically, an on-duty schedule is made up among the members and the on-duty person monitors the radios and self-dispatches according to department policy. This system seems to be popular in small volunteer departments.

- *Advantages:* The members are familiar with safety issues and have a forum to communicate. A pool of people to draw from is defined and available. ISO training is easily accomplished. An ISO network is formed.
- *Disadvantages:* Career (paid) departments may incur overtime expenses due to committee training, meetings, being on-call, and actual incident responses by members. Extra radios, pagers, and other ISO equipment must be either purchased for each member or passed from member to member according to the duty schedule.

All Eligible Officers Strong disciples of IMS use this method to delegate an ISO. In essence, the incident commander appoints an eligible officer to the ISO position. The eligibility level is established by the department based on which level of officer (lieutenant, captain, battalion chief, and so on) it feels is trained and experienced enough to be an effective ISO. To be proactive, some departments have the ISO system designed so that the third-arriving company officer or second-arriving battalion chief automatically reports to the IC as the ISO (if responding as a part of a crew, the officer's crew fills support tasks like running the accountability system, initiating crew rehab, or other like tasks). If this method is chosen, for it to be effective, the department *must* provide specific ISO training to each and every officer who may fill the ISO role.

- *Advantages:* The pool to draw from is as large as the department has companies or on-duty battalion chiefs. The officer filling the ISO role may have a crew to fill critical on-scene safety roles like rehab or running the crew accountability system. If an ISO is not needed by the IC, other assignments can be given or the officer is rotated back into staging as an available resource.

- *Disadvantages:* The department must ensure that all eligible officers are trained. This is specialized training for what could be a large group of officers.

Dedicated ISO   Of all the systems, this is perhaps the most desirable. The system amounts to a 24-hour duty position with an ISO always available. In large metropolitan areas, multiple ISOs can be on duty to cover the potential for simultaneous significant events.

- *Advantages:* The ISO is responding for a specific assignment and is positioned to have focused, duty-related training, experience, and proficiency. The ISO can assist with other health and safety assignments, documentation, preplanning, or proactive safety training when not actively engaged at an incident.

- *Disadvantages:* The system requires commitment from the department, including funding additional positions. Departments with a significant incident injury rate, a significant volume of working incidents, or a high potential for firefighter injury are the ones that typically choose this method.

The key to all these systems is to get a trained, recognized ISO on-scene in an automatic way—early.

HOW DOES THE ISO GET THE JOB DONE?

Define the ISO Plan

The whole point of a fire and rescue department's ISO plan is to ensure that the ISO is utilized at significant incidents. To get the job done, the ISO needs the department to define the options (many of which are covered in this book) and commit them to writing for utilization by the line officers. As the popular phrase says, "Don't Just Think it . . . Ink It! Appendix B includes sample SOPs for the design of an ISO system.

When a department does not formalize a system for automated or rapid delegation of the ISO function, or chooses not to, the ISO, if ever appointed, faces a difficult task. One area that has not been discussed is that of *authority* in designing an ISO system. What kinds of authority should the ISO have? This is an area of some controversy in the fire service ranks.

Issue: Should the ISO have the authority to stop an unsafe act and correct it on the spot?

Arguments:

- *Yes:* Those who agree that the ISO should be able to stop an unsafe act recognize that the whole purpose of the ISO is to make the incident safer. NFPA 1521 gives the ISO the authority to stop, alter, or terminate activities if an imminent threat exists.[5]
- *No:* Those who disagree believe that the ISO, by stopping or correcting any act on the fireground, is actually countermanding the incident commander. Acts deemed unsafe by the ISO should be immediately reported to the IC for a decision to stop, alter, allow, or correct them.

Regardless of the arguments, an important point is that the ISO can do tremendous damage to an effective program by exploiting the authority of the position. As with the other components of an effective ISO system, the authority issue needs to be addressed *prior* to the incident. Obviously, the intent of this book is to support the ISO's authority to stop, alter, or suspend operations that pose an imminent threat.

Train the ISO

As with any fire service discipline, a system for initial and ongoing training is essential. Many departments assume that the training/safety officer or any company officer should know what it takes to fill the ISO role. This assumption is dangerously flawed. Section Two of this book discusses some of the key training areas that help develop an ISO. Section Two makes it clear that the knowledge and skills required of an ISO exceed that of the typical fire officer I level. Combine the topical areas from Section Two with ongoing continuing education and the department has an ISO training program.

The Fire Department Safety Officers Association (FDSOA) can serve as a training resource for the ISO. The FDSOA Web site, www.FDSOA.org, offers network forums, products, and information on training opportunities to help the ISO. FDSOA is also accredited to facilitate professional qualification certification for the ISO.

Utilizing ISOs at drills, exercises, and other training sessions can serve as an opportunity to help keep them proficient.

Give the FDISO Tools to do the Job

There are a few tools that help the ISO make a difference in getting the job done. The following list is a summary of items that have been used to help ISOs around the country **(Figure 4-12)**.

Radio   As explained in subsequent chapters, the effective ISO is constantly roving and watching. A radio is essential to maintaining contact with the IC as well as monitoring the working crews.

Figure 4-12 *Essential incident safety officer tools include* at least *proper identification, radio, phone, documentation equipment, and flashlight.*

High Visibility   A florescent vest marked "Safety Officer," green helmet, or other unique identification pays dividends on the incident scene. Not only does this make the ISO easy to spot, but serves as a reminder to all crews that their safety is important. It's important to note that a unique identifier (like a vest of a different color than other command position vests) is the key. It's interesting that many ISOs report a remarkable change in firefighter behavior when the firefighter discovers that the "Safety Vest" is nearby. This may seem trivial—but it works. For example, the Philadelphia Fire Department uses a tan-colored vest for most command positions and a blue vest for Safety Officers. Green is a good color choice also because of its association with safety organizations (such as the National Safety Council or Fire Department Safety Officers Association).

Personal Protective Equipment (PPE)   To tour the scene, the ISO may have to cross into "hot zones" or other areas that require PPE. If assigned to check out interior or high-risk areas, the ISO should request a partner from the IC and be processed through the crew accountability system just like any crew. Also, the ISO sets a good example by utilizing appropriate PPE. When the ISO is wearing self-contained breathing apparatus (SCBA), again, a partner should be assigned. In these cases, the unique vest that the safety officer may be wearing becomes hidden. Some fire departments have fashioned or purchased identifying sleeves that slip over the SCBA cylinder.

Clipboard File Box   The ISO can make a difference if he or she is prepared for things the incident commander expects. A metal clipboard file box can give a working surface for notes, sketches, checklists, and other documents. The file box can contain accident reports, forms, quick reference sheets, required IMS forms, safety briefing forms and reminders, tablets, pencils, and other supplies. One idea worth considering: Keep a laminated sheet and grease pencil handy so quick signs can be made—like a "*Look up!*" sign that can be flashed to crews in a hot zone.

Miscellaneous   Cameras, stopwatches, barricade tape, gas detectors and monitors, whistles, electrical current detection instruments, cell phones, and other pieces of equipment have been suggested or used by many ISOs. These and other items may be necessary based on local needs.

Designing an ISO system may sound overwhelming, but know that there are many departments with good systems in place today. There is no need to reinvent the wheel! Hopefully, your department can use this chapter to outline a workable system to ensure that an ISO is available for significant incidents. Remember that the sooner the ISO position is filled, the better the chances are for a safe operation.

SUMMARY

The design and implementation of an effective incident safety officer system is just as important as having an ISO on scene. To ensure effectiveness, the ISO needs to be on-scene early. Therefore, it is best to preplan the ISO response. Certain incidents require the automatic response of an ISO, while others might reach a stage at which the incident commander needs to delegate the safety responsibility. In both cases, the procedures to get an ISO in place should be preplanned. Many factors influence the "who" part of assigning the ISO task. Advantages and disadvantages need to be discussed as a fire department determines who can best fill the ISO role. Once a system has been developed to ensure that an incident safety officer gets assigned early and often at working incidents, the individuals who fill the role should be given the training and tools to get the job done and to be effective.

REVIEW QUESTIONS

1. Explain the reasons that the ISO role should be preplanned.

2. List four examples of when an automatic ISO response is beneficial.

3. List four examples of when automatic ISO delegation should take place.

4. List three methods to get an ISO on-scene, and discuss the advantages and disadvantages of each.

5. Explain the authority given to the incident safety officer by NFPA standards.

6. List four tools that can help the ISO be effective on scene.

ADDITIONAL RESOURCES

Fire Department Safety Officers Association. Web site containing network forums, products, and training opportunity/certification information: Available at: www.FDSOA. org.

NOTES

1. NFPA 1521, *Fire Department Incident Safety Officer,* 2008 ed. (proposed) (Quincy, MA: NFPA, 2007), 6.1.2.

2. NFPA, *NFPA Journal,* 89, no. 4 (July/August 1995).

3. Centers for Disease Control & Prevention. Web site containing fatality reports: Available at: http://www.cdc.gov/niosh.

4. This generalization is based on the author's 15 years of traveling around the United States in support of incident safety officer training, research, and consulting.

5. NFPA 1521, *Standard for Fire Department Safety Officer* (proposed) (Quincy, MA: NFPA Publications, 2008).

Section

2

FRONT-LOADING YOUR KNOWLEDGE AND SKILLS

FRONT-LOADING AN INTELLIGENT AND SAFE FIREGROUND OPERATION

I am a battalion chief for a four-station career fire department near Seattle. Prior to shift work, I was assigned as chief of training. During my training division tenure I was determined to provide one full day of live-fire training to each and every firefighter each and every year. One of the objectives of this goal was to emphasize *strategic* issues. During my 25-year career, I had never participated in live-fire training that focused on strategic stuff.

- Problem identification (sizeup)
- Incident action planning
- Radio communication
- Incident management
- Personnel accountability

I wanted our fire officers to be *front-loaded,* that is, to know what an intelligent and safe fireground operation should look like and sound like. For example, fire officers were advised that a melted helmet was a reliable indicator of an unintelligent and unsafe fireground operation. I wanted to change the mind-set that a melted helmet was somehow "cool"; should a team emerge from the fire building with heat-damaged personal protective equipment, I wanted the team leader to be ashamed—not proud.

The problem was—and I believe this problem exists throughout North America—nobody has been front-loaded with what an intelligent and safe fireground operation *looks like* and *sounds like.*

It's easy to front-load the fireground with tactical and task-level training: stretching hose, climbing ladders, operating pumps, ventilation, operating nozzles, pumping, stokes rescue, searching, and forcible entry. Front-loading the fireground with strategic preparation is another matter. It is much easier to develop *great* tacticians than *good* strategists. My goal was (and still is) to develop great strategists.

Fast-forward a few years . . . We facilitated 18 to 22 regional multicompany live-fire exercises *every year for seven years.* Each day of live-fire training included four scenarios that incorporated five engines, one or two ladder trucks, a couple of chief officers, and an EMT ambulance. (Regional fire departments participated and shared the cost.)

Each scenario was run like an actual incident. Units were parked a block or two away, the incident was dispatched, apparatus arrival was sequenced, and officers were instructed to perform a thorough size-up, develop an action plan, communicate the plan, assign objectives, account for personnel, and so on. Chief officers established or assumed command and were required to manage the scenario from a formal command post.

At these "incidents," rarely was the front-loading at the tactical and task levels deficient. Generally, all the firefighters were task-level competent. The glaring problem that emerged, seen time and again, was the absence of strategic competence, and the lack of strategic competence is the result of inadequate fire officer front-loading.

We watched very competent fire officers and incident managers in action. Unfortunately, strategic competence was not the norm. When strategic deficiency was observed it was not because the fire officers were not smart people; it was because these smart people had never been front-loaded strategically.

Among our observations:

1. Many fire officers did not know how to perform a thorough fireground size-up.

2. Many fire officers did not know how to develop an incident action plan. (Some fire officers did not know they were supposed to.)

3. A lot of division and group supervisors had no idea how to supervise a division or group.

4. Radio traffic was frequently conversational and strategically meaningless.

5. Safety officers did not know what they were supposed to do.

6. Fire officers were often observed operating at task level.

7. Resources in staging were often overlooked or mismanaged.

8. Disturbingly, many participants considered live-fire training as an opportunity to melt their helmets.

Your fireground operations can only be as good as the front-load strategic investment made in your fire officers. Front-load solutions to the problems listed and you will develop competent fire officers. Bungled tasks do not kill firefighters; the absence of strategy kills firefighters. Make sure your fire officers—and your incident safety officers—have been strategically front-loaded so that they know what an intelligent and safe fireground operation looks like and sounds like.

Mark Emery, Battalion Chief
Woodinville Fire & Life Safety
King County, Washington

Chapter 5

PROFESSIONAL DEVELOPMENT AND MASTERY

Learning Objectives

Upon completion of this chapter, you should be able to:

- List the three areas that an ISO must "front-load" to help perform the functions of the ISO.
- Discuss the concept of "mastery" and its benefit to the ISO.
- Describe the relationships among knowledge, skill, and attitude.
- List the three components of an attitude.

INTRODUCTION

The first section of this book was focused on introducing the role of a fire department incident safety officer and presenting the basic concepts for utilizing the ISO. In this section we offer specifics on how you can prepare yourself to address the ISO's functions. As a starting place, the fire service has agreed that an ISO should meet the minimum requirements for Fire Officer 1, as outlined in NFPA Standard 1021: *Standard on Fire Officer Professional Qualifications.*[1] The Fire Officer 1 requirement, however, is only a starting point. Working ISOs around the country agree that, to be effective, they must acquire additional knowledge and skills well past the Fire Officer 1 level. Additionally, effective ISOs must acquire a certain *attitude* that helps them apply their acquired knowledge and skills. Your efforts to acquire the knowledge, skill, and attitude are what we will call "front-loading." ISO who spend the time front-loading are much better prepared to perform the multitude of challenges that they must face. In formal terms, front-loading is the first step towards professional development as an ISO.

MASTERY

mastery
the concept that an individual can achieve 90 percent of an objective 90 percent of the time

The goal of ISO professional development is to achieve "mastery." **Mastery** is defined as the ability of an individual to achieve 90 percent of an objective 90 percent of the time.[2] Some refer to this definition as the 90/90 rule. As it relates to the incident safety officer—and incident handling—mastery may seem hard to gauge. How do you "measure" your performance in such a way that you know you handled an incident with 90 percent efficiency and effectiveness? Perhaps a better way to describe mastery for the incident responder (and ISO) is the ability to perform with a certain "unconscious competence." That is, you perform your functions with an analytical automation that is deeply connected with and appropriate for the incident. Some argue that achieving mastery of the ISO function is not obtainable—a compelling argument, given the diversity of the emergency response world. The intent, however, is to encourage the pursuit of mastery—and therefore, ever increasing effectiveness and efficiency in filling the ISO role.

The words "efficiency" and "effectiveness" are used in many contexts and often interchangeably. The two words are often used without appreciation for what they imply. In this book, we use the word "efficiency" to mean doing things right. "Effectiveness" is used to reflect doing the right things.

Mastery Means Doing the Right Things Right		
Effectiveness +	**Efficiency** =	**Mastery**
(Doing the right things)	(Doing things right)	(Doing the right things right)

learning
the acquisition of knowledge, skills, and attitude to achieve mastery

performance
the demonstration of acquired mastery

To become efficient and effective, the ISO must learn, then perform. Learning and performance are not the same things. **Learning** is the acquisition of the knowledge, skills, and attitude needed to achieve mastery. **Performance** is the demonstration of acquired mastery in the incident environment. The two combine to make the ISO efficient and effective. Coming full circle, learning is the "front-loading" and performance is the application of that learning. Here in Section Two we look at the learning areas, and in Section Three we look at the performance side. In this chapter, the question is what knowledge, skills, and attitudes need to be learned?

ISO KNOWLEDGE

The list of knowledge topics for an ISO can be overwhelming. It may be better to look at some overarching subject areas as a start. For each subject area, the level of understanding needs to be, at minimum, Fire Officer 1 level:

- Building construction
- Risk/benefit concepts
- Fire behavior
- Firefighter physiology
- Hazardous energy
- Incident management systems

■ **Note**
Generally speaking, Fire Officer 1 level can be described as knowing the depth of the material and being able to bring the knowledge into application without supervision.

Generally speaking, Fire Officer 1 level can be described as knowing the depth of the material and being able to bring the knowledge into application without supervision. Each of the preceding topic areas deals with volumes of information and theory, and each warrants continual study and application if the ISO is to gain mastery. The key, however, is to make the transition from the book learning to the incident scene. Today, knowledge is the basis for skills and attitude. Previously, most firefighters earned their livelihood as a result of the skills they demonstrated. Today, rapidly changing technologies and environments require the fire service member to rely on knowledge to recognize the potential of an incident. The key is to *recognize* situations—and knowledge facilitates recognition.

ISO SKILLS

Most firefighters associate the word "skill" with motor tasks such as throwing ladders, performing an evolution, or manipulating a forcible entry tool. For the ISO, "skill" refers to intellectual tasks such as hazard reduction and problem solving. Skill objectives use adjectives like "determine," "predict," and "implement," whereas knowledge objectives use terms like "list," "describe," and "identify." From an incident commander's perspective, the ISO is expected to apply skills and offer judgment on many incident factors. The accompanying textbox (Essential ISO Skills) lists some of the incident factors and shows the skills an ISO must have to manage firefighter safety.

Essential ISO Skills

- *Fire behavior:* The ISO must be able to look at smoke conditions and determine the stage of the fire, the degree of fuel involvement, the structural impingement, the rate of heat release, and the potential for a hostile fire event. Additionally, the ISO needs to read smoke to determine if the firefighting effort is achieving positive results.

- *Building construction:* The ISO should have the *knowledge* to classify a building with regard to occupancy type, construction type, and materials used. Further, the ISO should possess the *skill* to predict weak points and collapse potential based on fire location, burn time, imposed loads, and resistance to loads. Additionally, the ISO should have the skill to establish appropriate collapse zones.

- *Physiology, kinesiology, and injury potential:* Can the ISO predict crew rehab needs, size up ergonomic stressors, and predict the injury potential for firefighters operating in an environment?

- *Risk/benefit thinking:* The ISO must decide if the action being implemented presents an acceptable risk to those taking it. The application of preestablished risk values can help the ISO determine acceptable risk.

As you can see, ISO skills make up an organized mental and physical activity. Sometimes, it involves only mental activity, but more often than not, it includes mental and physical activities. The primary issue, however, is organization. Skill development involves knowledge, sustained effort, and practice.

ISO ATTITUDE

Many fire departments are recognizing that it is not enough to concentrate on intellectual and motor skills. Attitudes seem to be receiving more attention in the training and education of fire service members. We know that attitudes

■ **Note**

To be ultimately effective and efficient, ISOs must use their acquired knowledge and skills to shape an attitude that supports the reduction of injury and death potential of firefighters.

■ **Note**

It would be counterproductive for ISOs to display such a discrepancy between what they say and what they do.

are learned and acquired gradually and, at times, incidentally. Attitudes can be shaped and altered by beliefs, values, and experiences. Given all this, it may seem difficult to change one's attitude. To be ultimately effective and efficient, ISOs must use their acquired knowledge and skills to shape an attitude that supports the reduction of injury and death potential of firefighters. This is easy to say—and perhaps more difficult to achieve. This entire book presents a certain attitude toward achieving the goal of preventing injuries and death.

Many firefighters *say* they have a positive safety attitude—but their actions suggest otherwise. It would be counterproductive for ISOs to display such a discrepancy between what they say and what they do. How can incident safety officers adopt and show a positive safety attitude? To start, they can acquire certain beliefs and values about injuries and deaths. The accompanying textbox (Positive Safety Attitude Values and Beliefs) lists some values and beliefs that can be viewed as consistent with a positive safety attitude.

Positive Safety Attitude Values and Beliefs

Values

Family first
Life preservation
Coworkers' well-being
Sacrifices that previous firefighters have made
Lessons learned from our past

Beliefs

Injuries are not an acceptable part of the firefighting profession.
Standards and laws are written to prevent future injuries and deaths.
Training and proficiency efforts are a daily commitment—and they never end.
Throwing resources at a fire is not always the solution.
Safety and self-discipline go together.

The next step in developing a positive safety attitude is founded on the understanding of the three components of attitude: knowledge, emotion, and action. The knowledge component deals with what a person knows—or doesn't know—about the topic. Emotion boils down to positive or negative feelings about the topic or, to use fire service language, what the person finds as acceptable or unacceptable. Action is the expression of the knowledge and emotion. Practically speaking, the ISO can view an action or a condition and nurture the proper attitude simply by asking:

- What do I know about this?

- How do I feel about it?

- How should I handle it to show a concern for safety?

These questions also work well to help you "check" your attitude. If you can apply this approach consistently, you are likely to display a uniform approach that will be viewed as a positive safety attitude. This may seem too simple—yet it is the essence of changing attitude.

The balance of this section helps you gain knowledge: the foundation of developing skills and acquiring a positive safety attitude.

SUMMARY

At minimum, ISOs must meet the professional qualifications as outlined in the NFPA 1021 requirements for Fire Officer I. This is a mere starting place. Effective and efficient ISOs develop knowledge and skills beyond the Fire Officer 1 level. Additionally, they must develop a positive safety attitude to be ultimately effective and efficient. The professional development effort to acquire ISO knowledge, skills, and attitude is called "front-loading." ISOs should strive for mastery (the 90/90 rule) in the performance of their functions. The acquisition of knowledge in the areas of building construction, risk/benefit concepts, fire behavior, firefighter physiology, hazardous energy, and incident management systems helps them recognize injury potential at incidents. For ISOs, "skill" refers to intellectual tasks—mostly analytical. Their analyses help ISOs form judgments that can facilitate managing firefighter safety. To be ultimately effective and efficient, ISOs must use the acquired knowledge and skills to shape an attitude that supports the reduction of injury and death potential of firefighters. Attitudes are comprised of knowledge, emotion, and action. ISOs are challenged with "checking" their attitude to ensure that they engender a positive, productive safety attitude.

KEY TERMS

Learning　The acquisition of knowledge, skills, and attitude to achieve mastery.

Mastery　The concept that an individual can achieve 90 percent of an objective 90 percent of the time.

Performance　The demonstration of acquired mastery.

POINTS TO PONDER

Firefighter Life Safety Initiatives

The National Fallen Firefighters Foundation (NFFF) hosted a first-of-its-kind Firefighter Life Safety Summit on March 10–11, 2004 in Tampa, Florida. The summit, consisting of more than two hundred fire and emergency service representatives

(*continued*)

(*continued*)

from over one hundred organizations and departments nationwide, was convened to support the United States Fire Administration's stated goal of reducing firefighter fatalities by 25 percent within five years and 50 percent within 10 years. After the event, the NFFF and the USFA released a report that details initiatives and recommendations for drastically reducing firefighter fatalities and injuries.

The Summit initial report identifies and provides additional background on 16 initiatives that were formulated by the summit participants:

1. Define and advocate the need for a cultural change in the fire service relating to safety, incorporating leadership, management, supervision, accountability, and personal responsibility.
2. Enhance the personal and organizational accountability for health and safety throughout the fire service.
3. Focus greater attention on the integration of risk management with incident management at all levels, including strategic, tactical, and planning responsibilities.
4. Empower all firefighters to stop unsafe practices.
5. Develop and implement national standards for training, qualifications, and certification (including regular recertification) that are equally applicable to all firefighters, based on the duties they are expected to perform.
6. Develop and implement national medical and physical fitness standards that are equally applicable to all firefighters, based on the duties they are expected to perform.
7. Create a national research agenda and data collection system that relates to the initiatives.
8. Utilize available technology wherever it can produce higher levels of health and safety.
9. Thoroughly investigate all firefighter fatalities, injuries, and near misses.
10. Ensure that grant programs support the implementation of safe practices and/or mandate safe practices as an eligibility requirement.
11. Develop and champion national standards for emergency response policies and procedures.
12. Develop and champion national protocols for response to violent incidents.
13. Provide firefighters and their families access to counseling and psychological support.
14. Provide public education with more resources and champion it as a critical fire and life safety program.
15. Strengthen advocacy for the enforcement of codes and the installation of home fire sprinklers.
16. Make safety a primary consideration in the design of apparatus and equipment.

For Discussion:

1. Which of the 16 firefighter life initiatives can be influenced by the actions of an incident safety officer?

2. For the initiatives that can be influenced by the ISO, which are tied primarily to a skill? Which are tied to attitude?

3. What do you feel are the some of the barriers to achieving the initiatives in your department?

4. What role does the ISO have in overcoming these barriers?

REVIEW QUESTIONS

1. What are three things that need to be acquired to front-load for the ISO function?

2. Discuss the concept of mastery and its benefit to the ISO.

3. How are efficiency and effectiveness different?

4. What is the essential difference between learning and performance?

5. Describe the relationships among knowledge, skill, and attitude.

6. What are the three components of an attitude?

7. To check your attitude, what three questions can be asked of yourself?

ADDITIONAL RESOURCES

Davies, Ivor K. *Instructional Technique*. New York: McGraw-Hill, 1981.

National Fallen Firefighters Foundation. *Firefighter Life Safety Initiatives:* Available through the U.S. Fire Administration at: http:// www.usfa.fema.gov.

NOTES

1. NFPA 1521, *Standard for Fire Department Safety Officer,* 2008 ed. (proposed) (Quincy, MA: NFPA Publications, 2008).

2. Ivor K. Davies, *Instructional Technique* (New York: McGraw-Hill, 1981).

Chapter 6

READING BUILDINGS

Learning Objectives

Upon completion of this chapter, you should be able to:

- Describe the relationship of loads and load imposition in a building.
- List the three types of forces created when loads are imposed on materials.
- Define columns, beams, and connections.
- Explain the effects of fire on building construction elements.
- List and define the five common types of building construction.
- Define and list several types of hybrid buildings.
- List, in order, the five-step analytical approach to predicting building collapse.
- List several factors that accelerate the time that a structural element will fail under fire conditions.

INTRODUCTION

ISOs must be able to give an incident commander explicit detail and their judgment regarding the collapse potential of a given building being attacked by fire. To make this judgment, they must draw from a significant knowledge base. Fire officers who draw from "experience" to predict building collapse are fooling themselves into a situation where a building will collapse without warning. Francis Brannigan, noted author of *Building Construction for the Fire Service,* says it best when he says, "Relying on experience alone is not sufficient. Firefighters must be aware of the theories and principles involved [in building construction]."[1] To say that a building collapses without warning is a flawed statement. The warning for structural collapse is in the ISO's ability to understand building construction and the effects of fire on the building. Granted, once collapse occurs, unpredictability increases because structural elements are no longer in a physical position to carry load as they were designed.

The Brannigan type of student is better prepared to make judgments regarding collapse potential. Further, ISOs who invest training time in conducting building surveys, construction site visitations, and preparing prefire plan reviews have a better knowledge base to pull from. The successful ISO becomes a lifelong student of building construction principles, materials, and collapse investigations.

Author's Perspective on Reading Buildings

I came into the fire service right out of high school. In high school, my "trades" experience was limited to working as a clerk in a small hardware store. Short of building a few childhood forts and teenage tree houses, I had no building construction knowledge or experience. My firefighter academy spent all of four hours on building construction—this in the 1970s. Many of the fire officers who were my instructors came from military or building trade backgrounds and assumed that every man knew how buildings were constructed.

In my third structure fire, I barely escaped a collapse. My crew and I dismissed the event as an acceptable danger of the job and shared a collective laugh at our good luck. Several years later, I was fortunate enough to attend a full-day class on steel construction by instructor Francis Brannigan. That class hit me like a ton of bricks (pun intended)! I discovered how little I really knew about building construction and got a glimpse of how much more I needed to learn about how fire affects structures. I became a forever student of building construction.

During my career, I used many sources to study building construction: all the books, Vincent Dunn's videos (deputy chief, retired, FDNY), and site visits. Although the books are important, the site visits became the most valuable tool in learning about buildings. As a company officer, I got myself in trouble for skipping daily assignments because I stopped by a construction site to learn the particulars
(continued)

(continued)

of the building. If another shift got a fire, I would go over the next day, look through the building, and see how it reacted. My dad was a fire investigator (for the Arvada Fire District (Colorado) and then later as a private fire investigator), and I used to help him "muck" on off-duty days just for the opportunity to see how the building reacted. As a training and safety officer, I continued to visit sites and borrow samples of new materials to see how they reacted under load in the burn house. If we got an acquired structure to burn, I continually monitored the structure between all the fun stuff. What fabulous learning!

Now that I'm "off the street," I continue to learn about buildings to help educate others on the street. The new materials and techniques used to construct and renovate buildings are simply amazing—yet very suspect to collapse under fire and heat conditions. I hope to continue my pursuit and share observations with you, but I must say that the best way to learn about buildings is to get out and *know your buildings!*

This ISO text is not a substitute for an in-depth study of the building construction books that detail the subject. This chapter, however, takes some essential building construction topics and thought processes and shows you how to evaluate and offer judgment on collapse potential (**Figure 6-1**).

Figure 6-1 *Building construction students apply their knowledge to better predict collapse.*

KEY TOPICS

Predicting building collapse is dependent on the application of essential building construction topics. ISOs use this knowledge base to help them "read" a building. Combined with smoke and fire observations, ISOs must apply a skill to predict collapse potential and collapse zones.

Imposition and Resistance of Loads

Obviously, buildings are constructed to provide a protected space to shield occupants from elements. The building must be built to resist the forces of wind, snow, rain, and gravity. Additionally, the intended use of the building can add a tremendous amount of weight, placing more stress on the building's ability to resist gravity. In building terms, these elements create *building loading*. Loads are then *imposed* on building materials. This imposition causes stress on the materials—called *force*. Forces must be delivered to the earth for the building to be structurally sound. The basis of all building construction techniques is to carry a load to earth. The load itself must be described or classified. The direction or application of a load to a component is the *imposition* of the load. The material that has a load imposed on it must resist the load. **Figures 6-2 and 6-3** show three ways loads are imposed and the forces they create in structures.

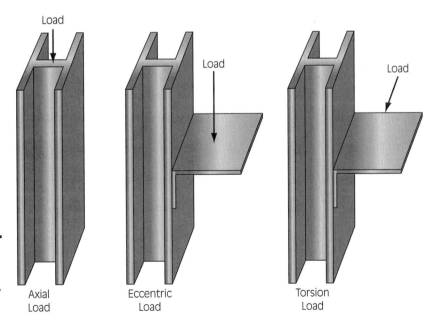

Figure 6-2 *Three types of loads can be transmitted through a structural member: axial, eccentric, and torsion.*

Axial
Load

Eccentric
Load

Torsion
Load

Application of Loads

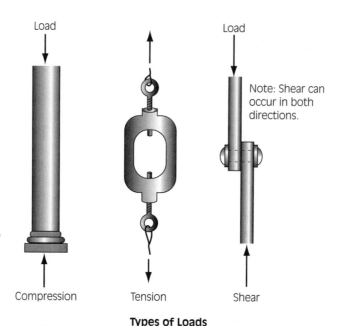

Load   Load

Note: Shear can
occur in both
directions.

Compression   Tension   Shear

Figure 6-3 *Loads are applied to a structural member as compression, tension, or shear forces.*

Types of Loads

Characteristics of Building Materials

The materials used to resist loads have load-bearing characteristics. Additionally, the fire service looks at how these materials react during a fire and how their ability to resist a load changes during fire conditions. Of importance is the dual concept of mass and fire resistance. The *mass* (or density) of a material directly impacts its fire resistance. In essence, mass *is* heat resistance, and heat resistance is time. From an ISO's point of view, the more mass a material has in a given surface area, the more time (or heat) is required before the material starts to degrade.

In the past, the fire service looked at the characteristics of four basic material types: wood, steel, concrete, and masonry. These materials can be found together or separately, and each reacts to fire in a different way (**Table 6-1**). Today, advanced material technologies have found their way into structural elements. Buildings are being assembled using plastics, graphites, wood derivatives, and other composites. In this section we cover the four basic building materials as well as some of the new composites.

Wood   Wood has marginal resistance to forces compared to its weight, but it does the job for most residential and small commercial buildings. We also know that wood burns and in doing so gives away its mass. The more mass a section of wood has, the more material must burn away before its strength is lost. This is true of native wood, that is, wood that has been cut from a tree.

TABLE 6-1   *Performance of common building materials under stress and fire.*

Material	Compression	Tension	Shear	Fire Exposure
Brick	Good	Poor	Poor	Fractures, spalls, crumbles
Masonry block	Good	Poor	Poor	Fractures, spalls
Concrete	Good	Poor	Poor	Spalls
Reinforced concrete	Good	Fair	Fair	Spalls
Stone	Good	Poor	Fair	Fractures, spalls
Wood	Good w/grain; poor across grain	Marginal	Poor	Burns, loss of material
Structural steel	Good	Good	Good	Softens, bends, loses strength
Cast iron*	Good	Poor	Poor	Fractures

*Some cast iron may be ornamental in nature and not part of the structure or load bearing.

engineered wood
products that consist of many pieces of native wood (chips, veneers, and saw dust) glued together to make a sheet, beam, or column

Engineered wood can react differently when exposed to heat from a fire. **Engineered wood** includes a host of products that consist of many pieces of native wood (chips, veneers, and sawdust) glued together to make a sheet, a long beam (trees grow only so tall!), or a strong column. The glues that bind wood products require only heat to break down. In some cases, the heated smoke from a distal fire can cause engineered wood products to degrade.

Steel   Steel is a mixture of carbon and iron ore heated and rolled into structural shapes to form elements for a building. Steel has excellent tensile, shear, and compressive strength. For this reason, steel is a popular choice for girders, lintels, cantilevered beams, and columns. Additionally, steel has high factory control; that is, it's easy to change its shape, increase its strength, and manipulate it during production. In a fire, steel loses strength as temperatures increase; the specific range of temperatures at which it loses strength depends on how the steel was manufactured. Cold drawn steel, like cables, bolts, rebar, and lightweight fasteners, loses 55 percent of its strength at 800 degrees Fahrenheit.

● **Caution**

Steel softens, elongates, and sags when heated, leading to collapse. Cooling structural steel with fire streams is just as important as attacking the fire!

Extruded structural steel used for beams and columns lose 50 percent of its strength at 1,100 degrees Fahrenheit. Structural steel also elongates or expands as temperatures rise. At 1,000 degrees Fahrenheit, a 100-foot long beam can elongate 10 inches! Now, imagine what that could do to a building. If a steel beam is fixed at two ends, it tries to expand and likely deforms, buckles, and collapses. If the beam sits in a pocket of a masonry wall, it stretches outward and places a shear force on the wall—which was designed only for a compressive force. The expansion of the beam can knock down the whole wall! Steel softens, elongates, and sags when heated, leading to collapse. Cooling structural steel with fire streams is just as important as attacking the fire!

Concrete   Concrete is a mixture of Portland cement, sand, gravel, and water that cures into a solid mass. Concrete has excellent compressive strength when cured. The curing process creates a chemical reaction that bonds the mixture to achieve strength. The final strength of concrete depends on the ratio of these materials—especially the ratio of water to Portland cement. Because concrete has poor tensile and shear strength, steel is added as reinforcement. Steel can be added to concrete in many ways. Concrete can be poured over steel rebar, which becomes part of the cured concrete mass. Cables can be placed through the plane of concrete and be tensioned, compressing the concrete to give it the required strength. Cables can be pretensioned (at a factory) or posttensioned (at the job site). Precast concrete consists of slabs of concrete that are poured at a factory and then shipped to a job site. Precast slabs are "tilted up" to form load-bearing walls; hence the phrase "tilt-up" construction.

All concrete contains some moisture and continues to absorb moisture (humidity) as it ages. When heated, this moisture content expands, causing the concrete to crack or spall. **Spalling** refers to a pocket of concrete that has basically crumbled into fine particles through the exposure to heat. Spalling can take away the critical mass of the concrete—that is, the mass used for strength. Reinforcing steel that becomes exposed to a fire can transmit heat within the concrete, causing catastrophic spalling and failure of the structural element. Unlike steel, concrete is a heat sink and tends to absorb and retain heat rather than conduct it. This heat is not easily reduced. Concrete can stay hot long after the fire is out, causing additional thermal stress to firefighters performing overhaul.

spalling
a pocket of concrete that has crumbled into fine particles through exposure to heat

Masonry   'Masonry' is a common term that refers to brick, concrete block, and stone. Masonry is used to form load-bearing walls because of its compressive strength, but it can also be used to build a veneer wall. A veneer wall supports only its own weight and is commonly used as a decorative finish. Masonry units (blocks, bricks, and stone) are held together using mortar. Mortar mixes are varied but usually contain a mixture of lime, Portland cement, water, and sand. These mixes have little to no tensile or shear strength; they rely on compressive forces to give a masonry wall strength. A lateral force that exceeds the compressive forces within a masonry wall causes its quick collapse.

● **Caution**

A lateral force that exceeds the compressive forces within a masonry wall causes its quick collapse.

Brick, concrete block, and stone have excellent fire-resistive qualities when taken individually. Many masonry walls are typically still standing after a fire has ravaged the interior of the building. Unfortunately, the mortar used to bond the masonry is subject to spalling, age deterioration, and washout. During a fire, masonry blocks (or bricks) can absorb more heat than the mortar used to bond them, creating different heat stresses that can crack the binding mortar. Whether from age, water, or fire, the loss of bond causes a masonry wall to be very unstable.

Composites   New material technologies have posed interesting challenges for the firefighting community. The term "composite" can be used for many things but in this case refers to a combination of the four basic materials, as well as various plastics, glues, and assembly materials. Of particular interest are the many wood products that are widely used for structural elements.

Lightweight wooden I beams (the slang term is "I joists") are a combination of two engineered wood products: laminated veneer lumber and oriented strand board. **Laminated veneer lumber (LVL)** is created by using sheet veneers of wood (with the same strand direction) that are glued and pressed together to form a piece of lumber. **Oriented strand board (OSB)** is sheeting created with wood chips (the strands are oriented in multiple directions) and an emulsified glue. LVL is used for the top and bottom chord of the I beam and OSB is used for the web between the chords (**Figure 6-4**). While the wooden I-beam is structurally strong (stronger than a comparable solid wood joist), it fails quickly when heated. Actually, no fire contact is required; ambient heating causes the failure of the binding glue and the potential for a quick collapse. As we know, the bottom of a beam is under tensile forces. If

laminated veneer lumber (LVL)
lumber created by gluing and pressing together sheet veneers of wood (in the same grain direction)

oriented strand board (OSB)
a wood sheeting consisting of wood chips (strands oriented in multiple directions) and an emulsified glue

Figure 6-4   *An engineered wooden I beam uses laminated veneer lumber (LVL) for the top and bottom chords and an oriented strand board (OSB) web. Heat alone can cause failure of the glues used within each type of material, as well as the glue that binds the two components.*

Figure 6-5 *Rapid heating of a composite truss can cause the metal to separate from the wood.*

the bottom of the beam falls off due to glue failure, the beam can immediately snap and collapse.

A new product known as **FiRP** (fiber-reinforced plastic, usually pronounced "ferp") is making its way into the construction industry. FiRP can be mixed with wood to give it incredible tensile strength—with minimal mass. As with most plastics, fire exposure can cause quick failure as the plastic melts. The term "composite" can also be used when using a mixture of common materials to form a structural member. Combining metal and wood to form a structural element can cause rapid collapse because the metal is likely to expand faster than the wood, causing a separating stress at the intersection of the two materials (**Figure 6-5**).

FiRP
fiber-reinforced plastic (usually pronounced "ferp")

Structural Elements

structural elements
building columns, beams, and connections

Buildings are an assembly of structural elements designed to transfer loads to the earth. **Structural elements** can be defined simply as columns, beams, and connections used to assemble a building. (A wall is nothing more than a long continuous column.) Each of these elements has its peculiar characteristics in how load is transferred. To predict collapse, the ISO must constantly evaluate structural elements to determine if they can still transfer the load as designed. If an element can no longer perform as designed, gravity takes over and pulls the element to earth.

Columns   Any structural component that transmits a compressive force parallel through its center is called a *column*. Columns can be formed as a wall or as a post, and they typically support beams and other columns (**Figure 6-6**).

Figure 6-6 *This column is supporting a beam, flooring, and another column. Columns are subject to compressive forces.*

Columns are typically viewed as the vertical supports of a building, even though they can be diagonal or even horizontal. The guiding principle is that a column is totally in compression.

Beams   A structural element that delivers loads perpendicularly to its imposed load is called a *beam*. Obviously, something must support the beam, and it is usually a column. It stands to reason that beams are used to create a covered space. In doing so, it is subjected to the imposed load of itself and of anything placed on it. This load causes the beam to deflect, that is, the top of the beam is subjected to a compressive force while the bottom of the beam is subjected to tension **(Figure 6-7)**. The distance between the top of the beam and the bottom of the beam dictates the amount of load the beam can carry or the distance the beam can span. The term "I beam" reflects the shape of the beam, viewed from either end. The top of the "I" is known as the *top chord;* the bottom of the "I" is the *bottom chord*. The material between the chords is known as the *web*. There are numerous types of beams although the principle method of load transfer remains the same. A few types of beams are as follows:

- *Simple beam:* Supported at the two points near its ends
- *Continuous beam:* Supported in three or more places
- *Cantilever beam:* Supported at only one end (or a beam that extends well past a support in such a way that the unsupported overhang places the top of the beam in tension and the bottom in compression)
- *Lintel:* Spans an opening in a load-bearing masonry wall, such as over a garage door opening (often called a "header" in street slang)
- *Girder:* Supports other beams

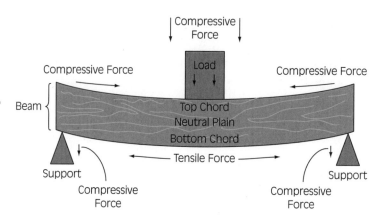

Figure 6-7 *A beam transfers a load perpendicularly to the load, creating compressive and tensile forces within itself.*

- *Joist:* A wood framing member used to support floors or roof decking (called a rafter when it is attached to a ridge board to form a peak)
- *Truss:* A series of triangles used to form a structural element to act as a beam(in many ways, a "fake" beam because it uses geometric shapes, lightweight materials, and assembly components to transfer loads just like a beam)
- *Perlin:* A series of beams placed perpendicularly to trusses or beams to help support roof decking

Connections   As mentioned previously, beams and columns (inclusive of walls) must be assembled in some fashion to effectively transfer loads. Connections provide the transfer. Often, the connection is the "weak link" in structural failure during fires. As can be imagined, the connection point is often a small, low-mass material that lacks the capacity to absorb much heat, thereby failing more quickly than an element with more mass, such as a beam or wall.

Connections are loaded in shear for the most part. There are three general types of connections: pinned, rigid, and gravity. *Pinned* connections use bolts, screws, nails, rivets, and similar devices to transfer load. In a *rigid* connection, the elements are bonded together such that the column (or load-bearing wall) is bonded to the beams. Reinforced concrete, welds, and glues are used to bond elements. *Gravity* connections are just that: The load from an element is held in place by gravity alone.

Together, structural elements defy gravity and make a building sound. A series of columns and beams used to hold up a building are often referred to as *skeletal frame* or *post and beam*. Beams resting on walls are simply called *wall-bearing* buildings. There are many ways to combine materials and structural elements to form a building that will carry loads. Over time, basic construction classifications have been developed to help us understand how a building is assembled.

CONSTRUCTION CLASSIFICATIONS

The general classification of construction methods helps the fire officer understand the materials, methods, and components present in a structure. There are many different ways to classify the construction of a building. Codes define building construction one way, contractors yet another. Even the fire service classifies construction styles differently within the profession. The important point regarding construction classification is not in the method chosen, but in your ability to look at a building and understand how the building is assembled and what materials were used. From this appreciation, you can apply your knowledge of material characteristics, load resistance, and special or weak features of the construction style and thereby be able to predict a collapse sequence.

The Five Types of Buildings

Over time, five broad categories of building construction types have been developed to help firefighters classify structures. These categories give the ISO a basic understanding of the arrangement of a building's structural elements and materials. Unfortunately, these broad classifications are dangerously incomplete for the ISO and may lead to deadly assumptions about the makeup of a building. For each building type, we outline its basic definition, its general configuration, and some associated historical fire spread problems. We also look at construction methods that do not fit into the five common types.

Type I: Fire-Resistive   In a Type I (Fire-Resistive) construction, the structural elements are of an approved noncombustible or limited combustible material with sufficient fire-resistive ratings to withstand the effects of fire and prevent its spread from story to story. Concrete-encased steel, monolithic-poured cement, and steel with spray-on fire protection coatings are typical of Type I. Most Type I buildings are large, multistoried structures with multiple exits. Fires are difficult to fight due to the large size of the building and the subsequent high fire load. Type I buildings rely on protective systems to rapidly detect and extinguish fires. If these systems do not contain the fire, a difficult firefight is required. Fire can spread from floor to floor on high-rises as windows break and the windows on the next floor up fail and allow the fire to jump. Fire can also make vertical runs through utility and elevator shafts. Regardless of how the fire spreads, firefighters are relying on the fire-resistive methods to protect the structure from collapsing. The collapse of fire resistive structures can be massive, as we saw in the World Trade Center collapse in New York City.

Type II: Noncombustible   In Type II (Noncombustible) construction, structural elements do not qualify for Type I construction and are of an approved noncombustible or limited combustible material with sufficient fire-resistive rating to withstand the effects of fire and prevent its spread from story to story. More often than not, Type II buildings are steel. Modern warehouses, small arenas, and newer churches and schools are built as noncombustible. Because the steel is not required to have significant fire-resistive coatings, Type II buildings are susceptible to steel deformation and resulting collapse. Fire spread in Type II buildings is influenced by the contents. While the structure itself does not "burn," rapid collapse is possible by means of the burning contents' heat release, stressing the steel.

Suburban strip malls, with concrete block load-bearing walls and steel truss roof structures, can be classified as Type II. Fires can spread from store to store through wall openings and shared ceiling and roof-support spaces. The roof structure is often of lightweight steel trusses that fail rapidly. More often than not, the fire-resistive device used to protect the roof structure is a dropped-in ceiling. Missing ceiling tiles, damaged drywall, and utility penetrations can render the steel unprotected. These buildings may have combustible attachments, such as facades and signs as well as significant content fire loading.

Type III: Ordinary "Ordinary" construction is often misapplied to wood frame buildings. By definition, ordinary construction includes buildings in which the load-bearing walls are noncombustible and the roof and floor assemblies are wood. Typically, this building is made of load-bearing brick or concrete block with wood roofs and floors. Ordinary construction is prevalent in most downtown or "main street" areas of established towns and villages. Firefighters have long called ordinary construction "taxpayers." This slang is derived from landlords who built buildings with shops and/or restaurants on the first floor with apartments above to maximize income to help pay property taxes. Newer Type III buildings include strip malls with block walls and wood truss roofs **(Figure 6-8)**.

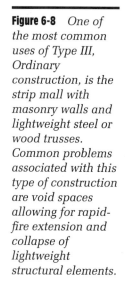

Figure 6-8 *One of the most common uses of Type III, Ordinary construction, is the strip mall with masonry walls and lightweight steel or wood trusses. Common problems associated with this type of construction are void spaces allowing for rapid-fire extension and collapse of lightweight structural elements.*

Ordinary construction presents many challenges to firefighters. In older buildings, numerous remodels, restorations, and repairs have created suspect wall stability and hidden dangers. Sagging or bowing load-bearing walls are often pulled back in alignment by tightening a steel rod that runs through the building from wall to wall, and a small interior fire can elongate this steel, causing catastrophic wall failure. These buildings can be spotted by decorative stars or ornaments (called *spreaders*) on the outside brick wall. Ordinary construction has many void spaces in which fire can spread undetected. Common hallways, utilities, and attic spaces can communicate fire rapidly. Masonry walls hold heat inside, making for difficult firefighting. Wood floors and roof beams are often gravity fit within the masonry walls, and these can release quickly and cause a general collapse, leaving an unsupported masonry wall.

Type IV: Heavy Timber   Type IV (Heavy Timber) buildings can be defined as those that have block or brick exterior load-bearing walls and interior structural members, roofs, floors, and arches of solid or laminated wood without concealed spaces. The minimum dimensions for structural wood is typically over 8 inches. Heavy Timber buildings, as the name suggests, are stout and are used as warehouses, manufacturing buildings, and some older churches. In many ways, a Type IV building is like a Type III, but larger-dimension lumber is used instead of common wood beams and trusses. Some firefighters mistakenly call Type IV buildings "mill construction." However, mill construction is a much more stout, collapse-resistive building that may or may not have block walls, and it is constructed without hidden voids. A new Type IV building is hard to find; the cost of large-dimension lumber and laminated wood beams makes this type of construction rare.

Fire spread in a Heavy Timber building can be fast due to wide-open areas and content exposure. The exposed timbers contribute to the fire load (heat release) of the fire. Because of the mass and large quantity of exposed structural wood, fires burn a long time. If the building housed machinery at one time, oil-soaked floors add heat to the fire and accelerate collapse. Once floors and roofs start to sag, heavy timber beams may release from the walls. The release is actually designed on purpose to help protect the load-bearing wall. This is accomplished by making a fire-cut on the beam, and the beam is gravity fit into a pocket within the exterior load-bearing masonry wall **(Figure 6-9)**. As the floor sags, it loses its contact point with the wall and simply slides out of its pocket without damage to the wall. However, because a masonry wall requires compressive weight from floors and roofs to make it sound, once the weight of floors and roofs is lost, the wall becomes an unstable cantilevered beam not designed for lateral loads like wind and fire streams.

Type V: Wood Frame   Wood Frame is perhaps the most common construction type; homes, newer small businesses, and even chain hotels are built primarily

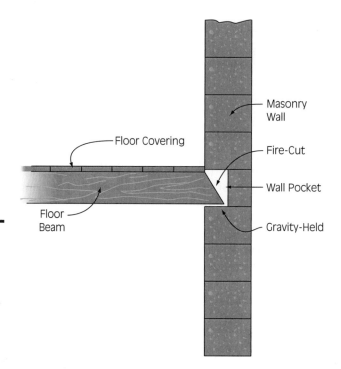

Figure 6-9 *Wood and heavy timber beams were often "fire-cut" so that a fire-damaged, sagging floor would simply slide out of the wall pocket to preserve the wall.*

of wood. Older wood frame buildings were built as *balloon frames,* that is, wooden studs run from the foundation to the roof, and floors are "hung" on the studs. Fire can enter the wall space and run straight to the attic. In the early 1950s, builders started using *platform framing,* in which one floor is built as a platform for the next floor. This creates fire stopping to help minimize fire spread. Newer wood frame buildings utilize lightweight wood trusses for roofs and floors; this is like a horizontal balloon-frame that can allow quicker lateral fire spread. Coupled with high surface-to-mass wood exposure, collapse becomes an early possibility. Some codes require truss spaces to have fire stopping every 500 square feet. Even with this fire-stopping feature, it remains dangerous to step onto (or under) the 500 square feet where the fire is! Wood frame structures may appear more like a Type III Ordinary building because of a brick wall appearance. Remember, brickwork may be a simple veneer to add aesthetics.

To protect structural members from a fire, wood frame construction typically uses gypsum board (drywall or sheetrock). Once finished, wood frame buildings typically have many "rooms" that can help compartmentalize content fires. Fire and heat that penetrate wall, floor, or attic spaces become a significant collapse threat, especially in newer buildings. Often, the only warning that fire has penetrated these spaces is the issuance of smoke from crawl space vents, gable-end vents, and eaves.

Other Construction Types (Hybrids)

Knowledge of only the five broad building types can actually lead to dangerous assumptions. Newer construction and alternative building methods may not fit exactly into one of the general types. Although not an official term, we call these **hybrid buildings**. Some buildings are actually two types of construction. **Figure 6-10** is an example of a restaurant that is built as a Type II Noncombustible, yet is topped with a large Type V Wood Frame structure to hide rooftop HVACs and cooking vent hoods. The square footage of the false dormers and wood frame structure exceeds that of most homes.

The advent of material technologies has produced construction types that are impressive from an engineering perspective. The fire service has very little research information on the stability of hybrid buildings during *actual* fires. One thought can be certain: Expect rapid collapse due to the low-mass, high-surface-to-mass exposure of structural elements! Let's examine some recent construction types that present interesting challenges to firefighters.

hybrid buildings
building construction methods that do not fit into the five classic building construction types; hybrids also include buildings built using more than one type of method

(A)   **(B)**

Figure 6-10   *(A) The decorative roof assembly is a Type V wood frame, whereas (B) the occupancy space is Type II Noncombustible.*

Lightweight Steel   New lightweight steel homes resemble wood frame homes. These buildings are actually a "post and beam" steel building with lightweight steel studs to help partition the home. OSB is added to the studs to help make the house more "stiff" and increase wind-load strength. Drywall protects the structural elements from compartment fires. The biggest collapse threat of these buildings comes with a fire in an unfinished basement—expect a rapid, and general collapse of flooring into the basement. The roof structure will likely sag before collapse as it is pinned together and uses steel perlins to help add rigidity.

● **Caution**

The ICF block wall is mostly EPS and will likely melt from the heat of a fire. ICF manufacturers emphasize that their products contain a fire-resistive additive. Remember that the ICF product may not contribute to the fire but will likely melt and decompose when exposed to other things that are burning. Expect rapid collapse of ICF block.

Insulated Concrete Forming (ICF)   ICF buildings use expanded polystyrene (EPS) to form a concrete mold for walls. Once the concrete cures, the EPS form is left in place as insulation. The two principal types are ICF block and ICF panel. ICF block use EPS blocks to build a wall (like a child's building blocks). Once the wall is up, concrete is poured into small cylinder cavities to give the wall

Figure 6-11   *In an ICF block wall, the thin circular openings are filled with high-slump concrete.*

compressive strength for the roof. ICF panel uses sheets, spaced apart, to form a mold similar to the typical basement foundation mold. Concrete is poured into the form. From the ISO point of view, the ICF panel wall includes significant concrete mass. The ICF block wall is mostly EPS and will likely melt from the heat of a fire. ICF manufacturers emphasize that their products contain a fire-resistive additive. Remember that the ICF product may not contribute to the fire but will likely melt and decompose when exposed to other things that are burning. Expect rapid collapse of ICF block. An example of ICF block is included in **Figure 6-11**.

SIP-wall (structural insulated panel)
a construction method that uses panels made from OSB and EPS for load-bearing walls and roofs; the panels are two sheets of OSB glued to both sides of an EPS sheet that is typically 6 to 8 inches thick

Structural Insulated Panels (SIP)　**SIP-wall** consists of panels made from OSB and EPS for load-bearing walls and roofs. The panels are simply two sheets of OSB glued to both sides of an EPS sheet that is typically 6 to 8 inches thick **(Figure 6-12)**. These panels are assembled like the house of playing cards you built as a child: Once the structures are finished, they may resemble a typical

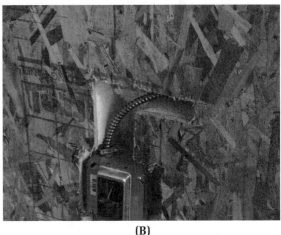

(A)　　　　　　　　　　　　　　　(B)

Figure 6-12　*(A) These are load-bearing SIP wall panels. Expanded polystyrene is sandwiched between OSB sheets. Fire can easily enter the wall space (B). Failure of the wall panel causes instability in the roof structure.*

wood frame building, but look for extended window and door jambs that indicate the wall is thicker than that of typical wood or masonry buildings. Heat alone can cause the interior EPS to contract (melt), leaving large combustible voids between the OSB. Fire can also enter the wall space through utility cuts and conduits. We do not have much fire experience with SIP-wall buildings. Simple analysis should lead a firefighter to expect rapid structural failure of these buildings during fires.

PREDICTING COLLAPSE

With a keen understanding of these topics, the incident safety officer can begin to analyze buildings during structure fires to establish if they will "behave" during firefighting operations in and around the structures. Obviously, the ISO must perform this analysis and make a judgment about collapse potential *before* the collapse occurs. A simple approach is presented here to provide a thought process for predicting collapse. As with most incident dynamics, nothing is absolute about predicting collapse. There is no perfect formula. For the ISO, however, the effort to analyze and predict collapse is absolutely essential to firefighter safety. The approach explained in this section can be applied to other structures such as bridges, cranes, mine shafts and wells, or any engineered "thing" attacked by fire. Building analysis during any incident should be cyclic, that is, performed on a regular basis as conditions change and time goes by. Using a five-step approach can help the ISO predict collapse.

■ Note
Building analysis during any incident should be cyclic, that is, performed on a regular basis as conditions change and time goes by.

The Five-Step Process for Predicting Collapse

Step 1: Classify the construction type.

Step 2: Determine structural involvement (read the smoke and flames).

Step 3: Visualize and trace loads.

Step 4: Evaluate time.

Step 5: Predict and communicate collapse potential (foundation for zoning).

Step 1: Classifying the Type of Construction

The easiest way to classify a building is to use standard language such as wood frame, ordinary, steel, and other familiar terms. Each classification type has strengths and weaknesses as well as fire spread patterns. By classifying the type of building, you can imagine how the materials and arrangement of structural elements might be impacted by fire and heat. If the building does not fit a classic type or is of multiple types, simply call it a hybrid. To help others understand the unique characteristics of the building, some ISOs like to further classify building types by the "era" they were built in. This provides a quick reference for all responders and tends to underscore the characteristics of the construction type. For example, it is one thing to call a building "Ordinary." It means more to fire officers if a building is a 1940s-built Ordinary versus a 2004-built Ordinary.

Step 2: Determining Structural Involvement

Determining whether a fire is a contents or structural fire is imperative **(Figure 6-13)**. Too often, fire departments use the term "structure fire" to dispatch crews. This may serve as a warning, but many incidents are merely content fires. Fire service leaders have suggested that crews should be dispatched to a "fire in a building." Only upon arrival and size-up is a fire classified as structural. For the sake of this book, "structural fire" means that the load-bearing components of a building are being attacked by fire or heat.

Fire in concealed spaces, content fires in unfinished basements, attic fires, heated exposed beam or truss, and the like are all examples of fires that have become "structural." The point is obvious: Once the structure is involved, the attention to possible collapse should be immediate. From the outside of a building, how would you know if the "hidden" structural elements are being attacked by fire or heat? The observation of fire coming from structural spaces should be obvious. Less obvious is what the smoke is doing (the subject of the next chapter). Without jumping too far ahead, know that dark gray or black smoke venting under pressure from structural seams, ridge boards, eaves, and attic vents is a significant indicator that heat and likely fire are present in that space. Also, unfinished wood that is being rapidly heated emits a brownish smoke.

Figure 6-13 *Once load-bearing structural members are attacked by fire, collapse may come quickly.*

Step 3: Visualizing and Tracing Loads

This step is more an art than a science. The ISO visually scans the building, tracing any load to the ground. In doing so, the ISO determines whether any structural element is carrying something it should not and whether key elements are being attacked by fire and/or heat. In visualizing all this, the ISO mentally "undresses" a building. This step helps the ISO define the weak link, which typically precipitates the collapse. Historical weak links are listed in the accompanying text box.

Weak Links That Lead to Collapse

Often, the collapse of a building is the result of a specific structural failure. This is the weak link. Each type of building construction has its own weak link; this is why study of building construction texts and site visits are so valuable.

Connections

Structural failure is often the result of a connection failure. Usually, the connection (bolt, pin, weld, screw, gusset plate, or the like) is made of a material with lower mass, or fire resistance, than the assemblies it connects. In some cases, the connection relies on gravity and an axial load to hold it in place; a shift or lateral load may cause the connection to fail.

Overloading

Excessive live and dead loads place stress on building components, which are vulnerable to early collapse when attacked by fire. In some cases, the building may not have been designed and built to handle excessive loads. Two feet of wet snow may have been considered when a flat roof building was built, but the addition of a second roof structure, storage, and/or rooftop signs—plus the two feet of snow—can overload the roof. Attic spaces are seldom designed to be used as storage areas, but the open space between trusses often lure the occupant into storing old files, parts, furniture, and other things. Remodeled buildings should always evoke a concern of overloading.

Occupancy Switch

Many buildings are used for occupancies they were never intended to support. For example, a lawyer moves his or her practice into a remodeled home: Walls are removed, increasing floor and roof spans to accommodate law books, legal files, heavy furniture, and office equipment. This massive weight addition leaves little room for fire tolerance and collapses quickly. Other examples include buildings constructed for a simple retail operation and converted to a manufacturing, storage, or medical use, which can bring excessive weight, higher fire loads, and alterations to the structure that were never meant in the original construction.

Trusses

Trusses are nothing more than "fake beams." Made of extremely light materials, trusses' geometric shape (triangles) and open space are used to replace a solid

(continued)

(*continued*)

beam. This is a deadly combination when fire and/or heat are introduced. Other questions arise when dealing with trusses, each indicating a weak link: Does the truss span a significant distance? What materials are being used to piece together the truss? Gusset plates typically hold together trusses. Gusset plates pop out quickly during rapid heating and fall off completely when a small portion of the truss woods burns away—after as little as 5 minutes of burn time. Trusses are used in roof, floor, and mezzanine assemblies.

Void Spaces

In Type III Ordinary construction, voids are numerous. Some voids may pass through masonry walls, causing fire spread from one store to the next in a row of buildings. The obvious collapse danger with void spaces is that the fire may be undetected with the simultaneous destruction of structural elements. Floor and roof truss spaces in lightweight wood construction create large combustible voids (usually packed with utility runs). Any signs of heat or fire from these spaces should trigger an immediate collapse warning.

Stairs

First-arriving firefighting crews rely on internal stairways to gain access for rescue and fire attack. For years, firefighters have found stairways to be durable and a bit stronger than other interior components. This is a dangerous assumption in newer wood frame buildings. Stairs are now being built off-site and simply hung in place using light metal strapping. Additionally, stairs are being made of lightweight engineered wood products that fail quickly when heated. Remember, press-glued wood chip products can fail from the heat of smoke—no flame is required!

Parapet Walls

A parapet wall is the extension of a wall past the top of the roof. Parapets are used to hide unsightly roof equipment and HVACs, and they give a building a "finished" look. Typically made of masonry materials, these walls are freestanding with little stability. Collapse may be caused by the sagging of the roof structure, which has the tendency to lift the parapet. Business owners hang signs, utility connections, and other loads on the parapet. During a fire, the steel cables and bolts holding these weaken and sag, placing additional eccentric load on the parapet, which accelerates collapse.

Step 4: Evaluating Time

Some structural elements fail as soon as fire (heat) reaches the material. Other materials absorb incredible heat for a long duration before they become susceptible to collapse. Time as a factor should be brought into the collapse equation, although the time it takes for gravity to overcome the structure during a fire is not predictable. A number of variables determine the amount of time a material can resist gravity and heat before failure. Factors that can accelerate the potential collapse time include:

- Low material mass, or high surface-to-mass ratio
- An imposed overload
- Higher BTU development (fire load)
- Alterations (undesigned loading)
- Age deterioration or the lack of care and maintenance of the structure
- Firefighting impact loads (fire stream force, accumulated water, forcible entry and ventilation efforts, weight of firefighting teams)
- Breakdown or loss of fire-resistive barriers

While no formula can predict collapse time, a few truisms can be applied regarding time (see text box).

Time Truisms for Predicting Collapse

- The lighter the structural elements, the faster it comes down.
- The heavier the imposed load, the faster it comes down.
- Wet (cooled) steel buys time.
- Gravity and time are constant; resistance is not.
- The time opportunity window for interior operations in lightweight buildings has been reduced.
- There is no time window for interior operations when a building is under construction, being renovated, or being disassembled.
- Brown or dark smoke coming from lightweight engineered wood products means that time is up.

NIOSH recently released an *Alert* publication to help prevent injuries and the deaths of firefighters due to truss system failures. The reason is included in one of the recommendations contained in the alert:

> Ensure that firefighters performing firefighting operations under or above trusses are evacuated as soon as it is determined that the trusses are exposed to fire—not according to a time limit.[2]

The message is clear: Do not rely on some "10-minute rule" or other such nonsense in trying to predict collapse. Many factors dictate the time it takes to trigger a collapse, and we should always predict the early collapse of trusses.

Step 5: Predicting and Communicating the Collapse Potential

Once the preceding steps are completed, the ISO can visualize a collapse scenario for the building involved in the incident. This information should be

collapse zone

areas that are exposed to trauma, debris, and/or thrust of a collapse; a collapse zone is a more specific form of a no-entry zone

Safety

The failure of a roof assembly may impose weight on interior partition walls that were not designed for the load. These become a "loaded gun," meaning that the slightest movement could cause an explosive release and collapse.

communicated to command in such away that includes the establishment of a collapse zone. A **collapse zone** consists of the areas that are exposed to trauma, debris, and/or thrust should a building or part of a building collapse. A collapse zone is a more specific form of a no-entry zone. Collapse zones should be considered no-entry zones for firefighters. Most fire service texts suggest that a collapse zone is at least 1½ times the height of the thing that is anticipated to fall. When it is not possible or desirable to honor the 1½ distance rule, a flanking approach can be considered when it is absolutely necessary to get closer. When flanking a collapse zone, firefighters should use spotters and have rapid withdrawal options communicated prior to the approach.

Typically, a building experiences a partial or general collapse. In a partial collapse, the building can accept the failure of a single component and still retain some strength. Often, a collapse is partial because another component picks up the weight of what fell. The partial collapse can also trigger a general collapse, in which the building cannot absorb the failure of an element and succumbs to gravity catastrophically. With new truss systems, interior floors can collapse into the basement, yet the walls and roof remain—a general collapse. Predicting collapse also includes determining whether walls will fall inward or out. Likewise, an inward fall of a wall may cause the bottom of the wall to kick out. The failure of a roof assembly may impose weight on interior partition walls that were not designed for the load. These become a "loaded gun," meaning that the slightest movement could cause an explosive release and collapse.

OTHER COLLAPSE CONCERNS

The ISO should use the analytical five-step process to predict and communicate collapse potential. In some situations, however, certain observations point to obvious collapse potential; let's call these "late signs." Although waiting for visual signs that a building will collapse is dangerous, especially in newer buildings, the following observations make collapse likelihood obvious:

- Deterioration of mortar joints and masonry
- Signs of building repair, including reinforcing cables and tie-rods
- Bulges and bowing of walls
- Sagging floors
- Abandoned buildings with missing roof, wall, or floor segments
- Large volumes of fire impinging on structural components and spaces
- Multiple fires in the same building or damage from previous fires

Buildings Under Construction

Buildings are especially unsafe during construction, remodeling, and restoration. The word "unsafe" applies not only to fire operations but also to rescues, odor investigations, and on-site inspections. Buildings need to meet fire and life safety codes only once they are completed. During construction, many of the protective features and fire-resistive components are incomplete. Additionally, stacked construction material may overload other structural components. This is not to say that contractors are using unsafe practices, but rather that exposed structural elements, incomplete assemblies, and material stacks all contribute to a rapid collapse if a fire develops. Defensive operations should be the default for all construction site fires.

Historical building restoration and general remodeling projects in buildings present hazards similar to those of buildings under construction. Firefighters may find the temporary shoring of walls, floors, and roofs while other structural components are being updated, replaced, or strengthened. Contractors may use simple two-by-fours to temporarily shore heavy timber, leading to disastrous results during fire conditions. The best approach for firefighters when responding to fires in buildings under construction is to be defensive. Once such hazards are detected, make sure everyone is out and accounted for, then attack the fire from a safe location. A building under construction can be replaced—a firefighter's life cannot.

SUMMARY

One of the important functions of an ISO is to offer judgment about the collapse potential of buildings during incidents. To do this, ISOs must front-load their building construction knowledge so that they can "read" the building and predict collapse potential. This ability comes from a long-term commitment to reading and studying building construction information. Knowledge of building construction starts with an understanding of the loads, forces, and materials found in the structural makeup of buildings. The ISO must also understand the effects of fires on materials and construction

types. The five classic types of construction are being challenged by new construction, or "hybrid," methods.

ISOs can use a five-step analytic process to predict collapse potential. The process is designed to factor in construction types, fire impingement, weak links, and time to arrive at a collapse potential. This potential must be communicated. Many factors determine when materials and construction design fail and gravity pulls down the building. Buildings under construction are "losers" from a firefighting point of view; they collapse quickly.

KEY TERMS

Collapse zone   Areas that are exposed to trauma, debris, and/or thrust of a collapse. A collapse zone is a more specific form of a no-entry zone.

Engineered wood   Products that consist of many pieces of native wood (chips, veneers, and saw dust) glued together to make a sheet, beam, or column.

FiRP   Fiber-reinforced plastic (usually pronounced "ferp").

Hybrid buildings   Building construction methods that do not fit into the five classic building construction types. Hybrids also include buildings built using more than one type of method.

Laminated veneer lumber (LVL)   Lumber created by gluing and pressing together sheet veneers of wood (in the same grain direction).

Oriented strand board (OSB)   A wood sheeting consisting of wood chips (strands oriented in multiple directions) and an emulsified glue.

SIP-wall (structural insulated panel)   A construction method that uses panels made from OSB and EPS for load-bearing walls and roofs. The panels are two sheets of OSB glued to both sides of an EPS sheet that is typically 6 to 8 inches thick.

Spalling   A pocket of concrete that has crumbled into fine particles through exposure to heat.

Structural elements   Building columns, beams, and connections.

POINTS TO PONDER

Glued Trusses

Much effort has been focused on informing firefighters of the dangers of the lightweight wood truss. Small-dimension lumber and gusset plates do not tolerate fire conditions well, leading to early collapse. Can you imagine ever wishing they still used gusset plates to hold a truss together? That time has come. Fully glued trusses are already being installed in floor and roof applications around the country (**Figure 6-14**). It is reported that glued trusses are actually stronger than a gusseted truss because there is more surface area being bonded.

Figure 6-14   *A glued floor truss.*

For Discussion:

1. Why do you think gusset plates are being replaced by glue in the building of trusses?

2. Given a fire or heat condition, what problems may arise for the glued truss?

3. What type of visual warning signs would you expect to see when the glued truss loses integrity?

4. Should the fire service differentiate between gusseted and glued trusses when preplanning or marking buildings? Why or why not?

REVIEW QUESTIONS

1. What are three ways loads are imposed on materials?

2. List the three types of forces created when loads are imposed on materials.

3. What is the definition of a beam?

4. Explain the effects of fire on steel structural elements.

5. How does a masonry wall achieve strength?

6. List and define the five common types of building construction. Give an example of each type that is located in your district or response area.

7. What is a hybrid building? List several types.

8. List, in order, the five-step analytical approach to predicting building collapse.

9. List several factors that accelerate the time that a structural element will fail under fire conditions.

ADDITIONAL RESOURCES

Angle, James A., David Harlow, Michael Gala, Craig Maciuba, and William Lombardo. *Firefighting Strategies and Tactics.* Clifton Park, NY: Delmar Publishers, a division of Thomson Learning, 2001.

Brannigan, Francis. *Building Construction for the Fire Service,* 3rd ed. Quincy, MA: National Fire Protection Association, 1992.

Building Construction for Fire Suppression Forces—Wood & Ordinary. Emmittsburg, MD: National Fire Academy, National Emergency Training Center, 1986.

Diamantes, David. *Fire Prevention: Inspection & Code Enforcement,* 2nd ed. Clifton Park, NY: Delmar Publishers, a division of Thomson Learning, 2003.

Dunn, Vincent. *Collapse of Burning Buildings.* Tulsa, OK: Fire Engineering Books and Video, a division of PennWell Publishing Company, 1988.

Frechette, Leon A. *Build Smarter with Alternative Materials.* Carlsbad, CA: Craftsman Book Company, 1999.

"High Rise Office Building Fire, One Meridian Plaza, Philadelphia, PA," Technical Report Series. Washington, DC: U.S. Fire Administration, 1991.

Morley, Michael. *Building with Structural Insulated Panels (SIPS).* Newtown, CT: Taunton Press, 2000.

NIOSH Alert, *Preventing Injuries and Deaths of Fire Fighters due to Truss System Failures.*

Publication No. 2005-1432. Cincinnati, OH: NIOSH Publications, 2005. Available at: www.cdc.gov/niosh.

"Without Warning, A Report on the Hotel Vendome Fire." Boston: Boston Sparks Association, n.d.

NOTES

1. Brannigan, Francis, *Building Construction for the Fire Service,* 3rd ed. (Quincy, MA: National Fire Protection Association, 1992).

2. NIOSH Alert, *Preventing Injuries and Deaths of Fire Fighters due to Truss System Failures* (Cincinnati, OH: NIOSH Publications, 2005). Publication No. 2005-132. Available at: www.cdc.gov/niosh.

Chapter 7

READING SMOKE

Learning Objectives

Upon completion of this chapter, you should be able to:

- Define "smoke."
- List common hostile fire events and their associated warning signs.
- List the four attributes of smoke.
- Describe what each of the four smoke attributes contributes to the understanding of fire behavior in a building.
- Define "black fire" and its relevance to firefighting efforts.
- Explain how influencing factors can affect smoke attributes.
- List the three steps in the reading smoke process.

INTRODUCTION

The current popular trend of "reading smoke" was triggered by Incident Safety Officer Academies that were developed by David Ross (Health and Safety Chief, Toronto Fire Services) and Dave Dodson for the Fire Department Safety Officer's Association in the 1990s. Though the concept of reading smoke is not new, prior to the ISO academies' effort, few had tried to define a process to assist ISOs in predicting fire behavior and to help them gauge fire-fighting progress. The practice of reading smoke has been around for many decades; fire officers handling America's fire epidemic in the seventies became quite proficient at the skill. Unfortunately, these sound tacticians felt that the ability to read smoke was based on experience and intuitiveness— and was difficult to teach. Further, the skills that these fire officers developed in reading smoke do not necessarily apply to today's fires. Low-mass synthetics and the consumer "glut" of the 1990s have led to a more volatile smoke and fire environment (more things to burn and more things made of plastic). This chapter presents the reading smoke process and offers suggestion on how an ISO can apply the process in a way to predict fire behavior and monitor fire attack effectiveness.

■ **Note**
Smoke leaving a structure has four key attributes: *volume*, *velocity (pressure)*, *density*, and *color*.

■ **Note**
Smoke can best be defined as the product of incomplete combustion that includes an aggregate of solids, aerosols, and fire gases that are toxic, flammable, and volatile.

smoke
the products of incomplete combustion, including an aggregate of solids, aerosols, and fire gases that are toxic, flammable, and volatile

"SMOKE" DEFINED

Reading smoke is not difficult, although for most fire officers, it takes an effort to break the "heavy smoke or light smoke" mentality that has come out of rapid size-up radio reports. Smoke leaving a structure has four key attributes: *volume, velocity (pressure), density*, and *color.* A comparative analysis of these attributes can help the ISO determine the size and location of the fire, the effectiveness of fire streams, as well as the potential for a hostile fire event like flashover. Before we can look at the meaning of each attribute, we must understand the science underlying what is seen in smoke.

In a simpler time, smoke was viewed as a by-product of incomplete combustion, specifically particulates (solids) that were suspended in a thermal column. Fire gases and aerosols were listed as separate products of the combustion process. In today's world, that oversimplification is dangerous. **Smoke** can best be defined as the products of incomplete combustion that includes an aggregate of solids, aerosols, and fire gases that are toxic, flammable, and volatile,[1] **(Figure 7-1)**. The solids suspended in a thermal plume include carbon (soot and ash), dust, and airborne fibers. Smoke aerosols include a whole host of hydrocarbons (oils and tar) and moisture. Fire gases are numerous, with carbon monoxide (CO), hydrogen cyanide (HCN), acrolein, hydrogen sulfide, and benzene leading the list. **Table 7-1** lists some of the more prevalent gases in smoke and their associated properties. The

Figure 7-1 *Smoke is an explosive aggregate of solids, aerosols, and gases. (Photo by Keith Muratori from FIREGROUND-IMAGES.com.)*

bottom line: Smoke is extremely flammable and ultimately dictates fire behavior in a building.

The ISO who focuses on the fire (flaming) to determine fire behavior is being set up for a sucker punch. To predict fire behavior in a building, the ISO needs to understand some realities that govern fire outcome. This understanding starts with the notion that "open flaming" is actually a good thing: The products of combustion are minimized because the burning process is more complete. In fact, the complete combustion of common materials renders heat, light, carbon dioxide, and water vapor. Within a building, the heat from flaming (exothermic energy) is absorbed in other materials that are not burning (contents and the walls/ceiling). These materials can only absorb so much heat before they start to break down and "off-gas" without flaming. At that point, smoke becomes flammable. Within a box (a room), the off-gassed smoke displaces air, leading to what is termed an underventilated fire. Underventilated fires do not allow the open flaming to complete a reaction with pure air, leading to increasing volumes of CO as well as the aforementioned

Table 7-1 *Properties of gases typically found in smoke.*

Gas	Flashpoint	Self-Ignition Temperature	Flammable Range in Air	Notes
Carbon monoxide (CO)	See notes	1123° F	12–74%	CO is considered a gas only and therefore does not have a flashpoint. The flammable range of CO is 12–74% at its ignition temperature. The flammable range of CO decreases below its ignition temperature. Below 300° F, the flammable range of CO is negligible.
Acrolein (C_3H_4O)	−15° F	450° F	3–31%	Acrolein is a by-product of the incomplete combustion of wood, wood products, and other cellulosic materials. Polyethylene can also render acrolein.
Benzene (C_6H_6)	12° F	928° F	1–8%	Most plastics release benzene while burning. Benzene is a common product of the burning of fuel oils.
Hydrogen Cyanide (HCN)	0° F	1,000° F	5–40%	HCN is produced when high temperatures break down nitrogen-containing products. HCN is quite flammable and is considered extremely toxic.

Source: Available at: http://www.bt.cdc.gov.

Safety

Two triggers may cause accumulated smoke to ignite: the right temperature and the right mixture.

smoke products.[2] Smoke is now "looking" to complete what was started. Two triggers may cause accumulated smoke to ignite: the right temperature and the right mixture.

Smoke gases that are below their ignition temperature need just a proper air mix and a sudden spark or flame to complete their ignition. Distal to the actual fire, a simple glowing ember or failing light bulb can spark an ignition. The ignition of smoke that has pressurized a room (or "box") likely results in an explosive surge. The ignition of accumulated smoke also changes the basic fire spread dynamics; instead of flame spread across surfaces of contents, the fire spreads with the smoke flow. The ISO who watches what the smoke is doing makes better decisions than the one focused on flaming, because the smoke tells you how intense the fire is about to become as opposed to how bad it currently is. John Mittendorf, noted author, instructor, and retired chief from Los Angles City Fire, said it best when he said, "Smoke is the fire talking to you."[3] Watching smoke flow can also warn the ISO that a hostile fire event is looming.

HOSTILE FIRE EVENTS

hostile fire event

an event that can catch firefighters off guard and endanger them: flashover, backdraft, smoke explosions, and rapid fire spread

Many believe that any uncontrolled fire in a building is a hostile fire event. This philosophy is right on. In this book, however, we use the phrase **hostile fire event** to describe the unique events that can catch firefighters off guard: flashover, backdraft, smoke explosions, and rapid fire spread. During interior firefighting operations, committed firefighters have restricted visibility and reduced ability to see the "big picture" of fire behavior. Similarly, the incident commander is busy with communications, status reports, tactical worksheets, and strategic planning at a fixed command post. The ISO is typically in a unique position to roam and watch smoke leaving the fire building. Given this flexibility, the ISO must watch smoke for warning signs that a hostile fire event is evolving. Historically, hostile fire events have been defined and argued by scientists, scholars, and fire experts. A review of fire event definitions in available texts and elsewhere shows that the debate continues.[4]

Table 7-2   *Hostile fire events.*

Event	Warning Signs	Notes
Flashover	• Turbulent smoke flow • Rollover • Autoignition outside	Flashover is an event triggered by radiant heat reflected by the "box." All gases reach their ignition temperature at virtually the same time due to rapid heat buildup in the box. If air is present, the box ignites explosively. If air is not present, the flashover is delayed until air is introduced. If smoke cannot exit the box, a stage is set for a backdraft.
Backdraft	• Yellowish-gray smoke • Bowing, black stained windows • Signs of extreme heat on outside of box or compartment	Backdraft occurs when oxygen is introduced into an environment where fire gases are above their ignition temperature and have been trapped in a box. Note: If sucking or puffing is witnessed near a box that is suspected of backdraft, the event is beginning. Sucking of air is a *late* sign of impending backdraft.
Smoke explosion	• Smoke that is being trapped above the fire • Signs of a growing fire • Signs of smoke starting to pressurize	A smoke explosion occurs when a spark or flame is introduced into trapped smoke that is below its ignition temperature but above its flashpoint. A proper air mix is necessary. Carbon monoxide is ignitable (with a spark or flame) around 300° F but has a small flammable range in air mix. As trapped CO heats up, its flammable range widens, making it easier to ignite with a spark or flame as it gets hotter.
Rapid fire spread	• Increase in smoke speed • Smoke flowing from hallways and stairways faster than a firefighter can move	Rapid fire spread occurs when smoke reaches sustaining temperatures that are above the fire point of prevalent gases. The gases can suddenly ignite when touched by an additional spark or flame. Fire spread changes from flame contact across content surfaces to fire spread through the smoke. This marks a significant change in fire spread behavior.

No single source can be cited to define hostile fire events or an agreed-on list of warning signs. Regardless, **Table 7-2** attempts to list the more common hostile fire events that trap and injure firefighters. Also included in the table are brief descriptions of each event and associated proactive warning signs. The ISO *must* study the warning signs of hostile fire events and watch for the signs as part of the reading smoke process.

Historically, firefighters are usually taught *reactive* warning signs for hostile fire events. For example, most firefighters list sudden heat buildup as a warning sign of flashover. Given the insulation provided by today's structural PPE ensemble, the sensation of heat is a dangerously late warning sign. Low ignition temperature gases are already ignitable when the firefighter feels heat! For this reason, firefighters—and the ISO—must learn about the *proactive* warning signs listed in Table 7-2.

> **! Safety**
> The ISO *must* study the warning signs of hostile fire events and watch for the signs as part of the reading smoke process.

VOLUME, VELOCITY, DENSITY, AND COLOR

The ISO must take a proactive approach by watching the four smoke attributes and determining the location, stage, and spread potential of a fire in a building, as well as the likelihood of a hostile fire event (**Figure 7-2**). Let's look at each of the attributes and how they contribute to fire behavior understanding.

Volume

Smoke volume by itself tells very little about a fire, but it sets the stage for understanding the amount of fuel that are off-gassing in a given space. A hot,

Figure 7-2
Comparing smoke volume, velocity, density, and color can help the ISO understand fire behavior. (Photo by Keith Muratori from FIREGROUND-IMAGES.com.)

clean-burning fire emits very little visible smoke, yet a hot, fast moving fire in an underventilated building shows a tremendous volume of smoke. Dampened material burns slowly and emits lots of smoke (typically of a lighter color). The changes in today's contents (low mass) can develop large volumes of smoke even though little flame is present. The volume of smoke can create an impression of the fire. For example, a small fast-food restaurant can be totally filled with smoke from a small fire. Conversely, it would take a significant event to fill the local big-box store with smoke. Once a container is full of smoke, pressure builds if adequate ventilation is not available. This fact helps us understand smoke velocity.

Velocity

The "speed" at which smoke leaves a building is referred to as *velocity*. In actuality, smoke velocity is an indicator of pressure that has built up in the building. From a tactical standpoint, the fire officer needs to know *what* has caused the smoke pressure. From a practical fire behavior point of view, only two things can cause smoke to pressurize in a building: heat or smoke volume. When smoke is leaving the building, its velocity is caused by *heat* if it rises and then slows gradually. Smoke caused by restricted volume immediately slows down and becomes balanced with outside airflow.

In addition to the speed of smoke, the ISO needs to look at its flow characteristic. If the characteristic is **turbulent smoke flow** (other descriptions may include "agitated," "boiling," or "angry" smoke), a flashover is likely to occur (**Figure 7-3**). Turbulent smoke flow is caused by the rapid molecular

turbulent smoke flow the movement of smoke through a building that is rapid and violent and that has expansive velocity (sometimes referred to as "agitated," "boiling," or "angry" smoke); indicates that the building (or compartment) cannot absorb more heat and is a precursor warning sign of flashover

■ **Note**

If the characteristic is turbulent smoke flow (other descriptions may include "agitated," "boiling," or "angry" smoke), a flashover is likely to occur.

Figure 7-3 *Turbulent smoke flow is a precursor to flashover. (Photo by Keith Muratori from FIREGROUND-IMAGES.com.)*

■ **Note**

If the box is still absorbing heat, the heat of the smoke is subsequently absorbed, leaving a more stable and smooth flow characteristic that is referred to as laminar smoke flow.

laminar smoke flow
the smooth and stable flow of smoke through a building; indicates that the building (or compartment) is still absorbing heat

🔔 **Safety**
The most important smoke observation is whether its flow is turbulent or laminar. Turbulent smoke is ready to ignite and indicates a flashover environment that may be delayed by improper air mix.

🔔 **Safety**
In essence, the thicker the smoke, the more spectacular the flashover or fire spread will be.

expansion of the gases in the smoke and the restriction of this expansion by the box (compartment). The expansion is caused by radiant heat feedback from the box itself; the box cannot absorb any more heat. This is the precursor to flashover. If the box is still absorbing heat, the heat of the smoke is subsequently absorbed, leaving a more stable and smooth flow characteristic that is referred to as **laminar smoke flow.** The word "laminar" is used to describe the smooth movement of a fluid, and smoke is basically fluid when moving through a building. The most important smoke observation is whether its flow is turbulent or laminar. Turbulent smoke is ready to ignite and indicates a flashover environment that may be delayed by improper air mix.

Comparing the velocity of smoke at different openings of the building can help the fire officer determine the location of the fire: Faster smoke is closer to the fire seat. Remember, however, that the smoke velocity you see outside the building is ultimately determined by the size and restrictiveness of the exhaust opening. Smoke follows the path of least resistance and loses velocity as the distance from the fire increases. To find the location of fire by comparing velocities, you must compare only like-resistive openings (doors to doors, cracks to cracks, and so on). More specifically, you should compare cracks in a wall to other cracks in walls. Do not compare the speed of smoke leaving a crack in window glass to the cracks in a wall; the resistance to smoke flow is different. There are some shortcuts to finding the location of the fire by reading velocity. We talk about those after we address density and color.

Density

While velocity can help you understand much about a fire (how hot it is and where), density tells you how bad things are going to be. The density of smoke refers to its thickness. Since smoke is *fuel*—containing airborne solids, aerosols, and gases that are capable of further burning—the thickness of the smoke tells you how *much* fuel is laden in the smoke. In essence, the thicker the smoke, the more spectacular the flashover or fire spread will be. Smoke thickness also sets up fuel continuity. On a practical basis, thick smoke spreads a fire event (like flashover) farther than less dense smoke. Even though we take turbulent smoke as a flashover warning sign, thick, laminar-flowing smoke can ignite because of the continuity of the fuel to the source.

One other point regarding smoke density: Thick, black smoke in a compartment reduces the chance of life sustainability due to smoke toxicology. A few breaths of thick, black smoke renders a victim unconscious and causes death in minutes. Further, the firefighter crawling through zero-visibility smoke is actually crawling through ignitable fuel. Modern fire tests are showing that smoke cloud ignition can happen at lower temperatures than fires of even ten years ago.[5] We can thank plastics and low-mass materials for making smoke more explosive than ever.

Color

Most fire service curriculums teach us that smoke color indicates the "type" of material that is burning. In reality, this is true for single-fuel or single-commodity fires. In typical residential and commercial fires, it is rare that a single fuel source is emitting smoke; the smoke seen leaving a building is a mix of colors. For the ISO, smoke color tells the stage of heating and points to the location of the fire in a building. Virtually all solid materials emit a white "smoke" when first heated. This white smoke is mostly moisture. As a material dries out and breaks down, the color of the smoke darkens. Wood materials change to tan or brown whereas plastics and painted or stained surfaces emit a gray smoke. Gray smoke is a result of moisture mixing with carbons and hydrocarbons (black smoke). As materials are further heated, the smoke leaving the material eventually is all black. When flames touch surfaces that are not burning, they off-gas black smoke almost immediately. Therefore, the more black the smoke you see, the hotter the smoke is. Black smoke that is high velocity and very thin (low density) is flame-pushed. Interpreted, thin, black smoke means that open (and ventilated) flaming is nearby.

Smoke color can also tell you the distance to a fire. As smoke leaves an ignited fuel, it heats up other materials and the moisture from those objects can cause black smoke to turn gray or even white over distance. As smoke travels, carbon content from the smoke is deposited along surfaces and objects, eventually lightening the smoke color. So the question is whether white smoke is a result of early-stage heating or of late-stage heating smoke that has traveled some distance? The answer is to look at the velocity. Fast moving white smoke indicates that the smoke you see has traveled some distance. White smoke that is slow or lazy is most likely indicative of early-stage heating.

One more important note about smoke color—namely, brown smoke. Unfinished wood gives off a distinctive brown smoke as it approaches late-stage heating (just prior to flaming). In many cases, the only unfinished wood in a structure is what makes up the wall studs, floor joists, and roof rafters and trusses. Brown smoke from structural spaces indicates that the fire is transitioning from a contents fire to a structural fire **(Figure 7-4)**. Using our knowledge of building construction—especially lightweight structural components and gusset plates—the issuance of brown smoke from gable-end vents, eaves, and floor seams is a warning sign of impending collapse. Remember that engineered wood products like OSB and LVL lose strength when heated. The glues of these products break down with heat and do not necessarily need flames to come apart.[6] Brown smoke from structural spaces containing glued trusses, OSB, or LVL can indicate that critical strength has been already lost.

Knowing the meaning of smoke attributes helps the ISO paint a picture of the fire. By combining the attributes observed, the ISO creates a mental picture of the fire. Compare smoke velocity and color from various openings to locate the fire. Faster and/or darker smoke is closer to the fire seat, whereas

■ **Note**

For the ISO, smoke color tells the stage of heating and points to the location of the fire in a building.

Safety

● Brown smoke from structural spaces indicates that the fire is transitioning from a contents fire to a structural fire.

Safety

● Brown smoke from structural spaces containing glued trusses, OSB, or LVL can indicate that critical strength has been already lost.

Figure 7-4 *Brown smoke from structural spaces indicates that unfinished wood is being heated.*

slower and/or lighter smoke is farther away. Typically you see distinct differences in velocity and colors from various openings. When the smoke appears uniform—that is, it is the same color and velocity from multiple openings—you should start thinking that the fire is in a concealed space (or deep-seated) (**Figure 7-5**). In these cases, the smoke has traveled some distance or has been

Figure 7-5 *Smoke that appears the same color and velocity from multiple openings indicates a deep-seated fire. (Photo by Keith Muratori from FIREGROUND-IMAGES.com.)*

pressure-forced through closed doors or seams (walls and concealed spaces), which "neutralizes" color and velocity prior to exiting the building. Upon seeing smoke that is the same color and velocity being pushed from multiple building seams, the ISO should inform the IC that the fire may have extended to concealed spaces.

A Smoke Reading Shortcut from the Street

Through the 1980s and 1990s, I spent time doing professional fire department visitations all over the United States. During theses "ride-alongs," I was fortunate to listen and watch, absorbing useful information from veteran fire officers at working fires. I was consumed with learning the little tricks shared by these veterans. Some of the smoke reading processes presented in this chapter come directly from these visitations.

One of my favorite tricks was shared by a veteran commander of hundreds of fires in a large metropolitan fire department. He told me that the fastest way of figuring out where a fire is in a building is to watch for the fastest smoke from the most resistive opening. After hearing this shortcut, I tried not only to practice it but to prove it wrong. In my own experience, I've found this to be a pretty accurate shortcut. Over time, I've altered the shortcut by adding color to the equation: *Watch for the fastest/darkest smoke from the most resistive crack.*

The rationale is that smoke leaving a building through a highly resistive crack has lost most of its heat energy and carbon/hydrocarbon color as it filters through the crack. If the smoke still has darkness and speed after leaving the resistive crack, it must be superhot and close to a fire source.

I've shared this shortcut with many fire officers when teaching "The Art of Reading Smoke" classes, and many take the time to e-mail me tales of how this shortcut helped in finding the fire.

Dave Dodson

Black Fire

black fire
a slang term used to describe high-volume, turbulent, ultradense, and deep-black smoke; a sure sign of impending flashover

Black fire is a slang term that fire officers have used for years. While hardly scientific, it is a good phrase to describe smoke that is high-volume, has turbulent velocity, is ultradense, and is deep black. Black fire is a sure sign of impending autoignition and flashover (look at Figure 7-3 again for an example of black fire). In actuality, the phrase "black fire" is accurate—the smoke itself is doing all the destruction that flames would cause: charring, heat damage to steel, content destruction, and victim death. Black fire can reach temperatures of over 1,000 degrees Fahrenheit! The ISO should report black fire conditions, and no firefighter should be in or near compartments emitting black fire. The solution to black fire is the same as that for flames: Vent and cool!

OTHER FACTORS THAT INFLUENCE SMOKE

Weather, thermal balance, container size, and firefighting efforts can change the appearance of smoke. Let's look how each of these can influence smoke.

Weather

It is important to know how weather affects smoke. Once smoke leaves a building, the outside weather can influence its appearance. Temperature, humidity, and wind change the look of the smoke. Virtually every element of smoke is heavier than air, yet it rises due to heat. If the outside air is cooler than the smoke, the smoke should rise until the outside air cools it. It stands to reason that cold air temperatures cool smoke faster and cause it to stall and/or fall. A hot day finds the smoke rising much farther because cooling is more difficult. Humidity in the air increases resistance to smoke movement by raising air density. Low humidity allows smoke to dissipate more easily. Humid climates keep the smoke plume tight. When outside air temperatures are below freezing, hot smoke leaving the building turns white almost instantly. This phenomenon is the result of smoke moisture or humidity condensing. Hot, dry smoke contacting cold, humid air changes to white. The opposite is also true: Hot, moist smoke contacting cold, dry air turns white. Practically speaking, the effects of wind on smoke should be obvious and should be taken into account when viewing the smoke leaving a building. Wind can even keep smoke from leaving an opening. Wind can rapidly thin and dissipate smoke, making it difficult to fully view its velocity and density. In a well ventilated building, wind can speed up smoke velocity and give a false read on heat or location, although it should fan flaming. Firefighters engaged in an interior fire attack downwind of a wind-fed fire are in danger of being overrun by the fire! If the weather is cold and humid, the smoke should sink rather quickly and stay dense; yet you arrive and see that the smoke is rising straight up and not cooling very fast. This indicates an exceptionally hot fire. Inversely, a hot, humid day should find the smoke climbing straight up into the atmosphere; yet you see that the smoke is coming out and lying down. This indicates that the smoke has been cooled, either by distance (a deep-seated fire) or by a sprinkler system. If the building is sprinkled, then the presence of low-lying smoke can indicate that the fire is not being controlled by the system.

Thermal Balance

Most buildings do not allow fires to maintain thermal balance. Simply stated, thermal balance is the notion that heated smoke rises and in doing so creates a draft of cool air into the flame (heat) source. As more air is drafted into the flame, the fire should grow. As the fire grows, it consumes the oxygen within the air, creating more of a draw of air. Ceilings, windows, doors, and inadequate airflow disrupt thermal balance. If a fire cannot draft and draw air, it soon drafts and draws smoke, thereby choking itself, leading to more incomplete combustion, and creating denser smoke. As explained, the thicker the smoke is, the more explosive it becomes. Viewed from outside a building, a fire out of thermal balance shows signs that air is being "sucked" through the smoke. Sucking, puffing, and "breathing" signs indicate that a fire is out of thermal balance. From the ISO's perspective, signs of air being sucked into a building indicate that the fire is intense, yet struggling for proper airflow. A sudden inflow of air can cause the fire to take off—trapping firefighters.

Container Size

All smoke observations must be analyzed in proportion to the building. For example, smoke that is low volume, slow velocity, very thin, and light colored may indicate a small fire, but only if the building (or "box") is small. The same smoke attributes from several openings of a big-box store or large warehouse can indicate a large, dangerous fire. Historically, firefighters have been killed at fires that were reported as "light-smoke showing." Remember, the size of the building is an important indicator of the significance of the smoke leaving. Light, thin smoke showing from more than one opening of a very large building is a significant observation.

Firefighting Efforts

The ISO is usually in the best position to tell whether firefighting efforts are being successful. By watching the smoke outside a building, the ISO can determine the effectiveness of the firefight. All four attributes of smoke should change in a positive, continuous manner if fire stream and ventilation efforts are appropriate. Smoke volume should rise (as steam displaces smoke). Contents that were flaming should start "smoking white" as they are cooled, adding more smoke to the mix. Smoke velocity initially surges as steam expands but should gradually slow as heat is reduced in the building. Turbulent smoke velocity should change to laminar flow. Smoke density should thin. Obviously, the smoke color should eventually turn to pure white. If all four attributes are not changing quickly (as described here), the ISO should judge the firefighting efforts as insufficient and share the observations with the IC.

Forced-ventilation tactics should cause an increase in smoke velocity. When ventilation fans are being used, the ISO should watch the smoke for signs that the fans are helping, rather than hurting, the fire suppression efforts. If smoke becomes darker and thicker, the fan is actually making conditions worse. The advent of positive pressure ventilation (PPV) tactics for fire attack has gained in popularity. Occasionally, the use of PPV can lead to disastrous results if not used properly. PPV as a fire attack tactic is contraindicated if smoke is turbulent (a flashover warning sign). PPV is also contraindicated if the location of the fire seat is unknown or if the fire is suspected to be in a vented, combustible void space. Again, if smoke becomes thicker and darker when PPV is being used, the situation is getting worse.

READING SMOKE: THE THREE-STEP PROCESS

The ISO may view the reading smoke principles as complicated or time-consuming. However, once you capture the basics and start practicing them, your ability to read smoke improves exponentially and you can read smoke in seconds! The principles can be incorporated into a process for rapid application. By following three simple steps, the ISO can refine and rapidly apply the principles.

Step 1: View the volume, velocity, density, and color of smoke; then compare the differences in the attributes from each opening from which smoke is emitting. This exercise should give you a strong understanding of the fire size, location, and spread potential, and it allows you to capture any warning signs of hostile fire events.

Step 2: Analyze the contributing factors to determine if they are affecting volume, velocity, density, or color. Remember, the size of the box can change the meaning of volume, velocity, density, and color. Likewise, weather and firefighting efforts can change each attribute. This step should refine and/or confirm your read on fire behavior and firefighting effectiveness.

Step 3: Determine the rate of change of each attribute. As you watch smoke, each attribute may be changing before your eyes. If the rate of attribute deterioration can be measured in seconds, it is likely that firefighters will be trapped or injured by fire spread. In other words, the firefighters are at risk.

These three steps should give ISOs a good understanding of the fire and help them predict what the fire will do next. If a hostile fire event appears imminent, immediate action must be initiated to prevent firefighter injury or death. As firefighting efforts progress over time, look for positive changes in the smoke conditions. **Table 7-3** outlines some of the shortcuts presented in

TABLE 7-3 *Reading smoke shortcuts.*

What You See	What It Can Mean
Black with push	Close to the seat of the fire or superhot smoke capable of instant ignition
White smoke with velocity	Heat-pushed smoke that has traveled a distance or has had the carbon filtered (like smoke through a crack)
Same color (white/gray) and same velocity from multiple openings	Deep-seated fire, possibly located well within a building or in combustible voids and concealed spaces
Brown smoke	Unfinished wood reaching late heating (can support flame); usually a sign that a contents fire is transitioning into a structural fire; when coming from structural spaces of lightweight wood structures, a warning sign of collapse
Turbulent smoke	Warning sign of impending flashover
Yellowish-gray smoke from cracks or seams	Warning sign of impending backdraft
Thin, black smoke moving fast	Flame-pushed smoke; nearby fire that is well ventilated
Smoke moving faster than firefighters can crawl	Warning sign that rapid fire spread is imminent

this chapter. Remember, however, that shortcuts are just that—quick ways of summarizing what you see. Nothing is absolute about reading smoke. Smoke, just like fire, is dynamic and is influenced by many variables.

SUMMARY

The ISO is in a unique position to observe and predict fire behavior by watching smoke exiting a building. Predicting fire behavior is founded on an understanding of the physical and chemical properties of smoke as well as on the specific warning signs of impending hostile events. Hostile fire events include flashover, backdraft, smoke explosion, and rapid fire spread. Most firefighters know the reactive warning signs for hostile fire events, but the ISO, as well as all firefighters, should learn the proactive warning signs of hostile fire events. In recent years, plastics and low-mass contents have made smoke more explosive. Smoke volume, velocity, density, and color can be compared to paint a picture of the fire in a building. Dynamic factors such as weather, container size, thermal balance, and firefighting efforts all influence the appearance and behavior of smoke. The ISO should also judge the rate at which smoke attributes are changing. With this understanding, the ISO can apply a process of reading smoke. Communicating smoke observations can help change action plans or warn firefighters of hostile fire events and therefore prevent firefighter injuries and death.

KEY TERMS

Black fire   A slang term used to describe high-volume, turbulent, ultradense, and deep-black smoke; a sure sign of impending flashover.

Hostile fire event   An event that can catch firefighters off guard and endanger them: flashover, backdraft, smoke explosions, and rapid fire spread.

Laminar smoke flow   The smooth and stable flow of smoke through a building; indicates that the building (or compartment) is still absorbing heat.

Smoke   The products of incomplete combustion, including an aggregate of solids, aerosols, and fire gases that are toxic, flammable, and volatile.

Turbulent smoke flow   The movement of smoke through a building that is rapid and violent and that has expansive velocity (sometimes referred to as "agitated," "boiling," or "angry" smoke); indicates that the building (or compartment) cannot absorb more heat and is a precursor warning sign of flashover.

POINTS TO PONDER

Smoke from Hell

The evidence is compelling. The building fires you respond to today and the ones you'll respond to tomorrow are producing some of the most explosive, toxic, and volatile smoke that we, as a fire service have ever experienced. Thanks to the work of the National Institute of Standards and Technology (NIST), the Bureau of Alcohol, Tobacco, and Firearms (ATF), and numerous product safety test laboratories, we are beginning to document the "new" building fire environment. Many have already claimed that the fire is the least of your problems—it's the smoke coming from low-mass synthetics that are being heated without burning (pyrolysis). According to some research, smoke production is 500 times that of the "wood-based" fires we saw a few decades ago.

The list of chemical products found in the smoke of a "typical" residential fire will challenge the imagination of most chemists: Hydrogen cyanide, polyvinyl chloride, hydrogen sulfide, hydrogen chloride, polynuclear aromatic hydrocarbons, formaldehyde, phosgene, benzene, nitrogen dioxide, ammonia, phenol, acrolein, and halogen acids. Further, the chemical decomposition process of pyrolysis is engaging the mechanisms of chain scission, chain stripping, and cross-linking (think rocket-science-like chemistry). It gets worse. We used to measure the typical compartment (room) fire time-temperature curve using a ten-minute cycle. Recent research tells us that the cycle is closer to five minutes.

Now we'll get personal. A review of firefighter news items, investigative reports, and incident data indicates that:

- Firefighter cancer rates are escalating.
- Firefighters that get just a few breaths of zero-visibility smoke either die or face significant medical issues that can become chronic.

(continued)

(*continued*)
- Firefighters are being treated for cyanide poisoning from just a "little" smoke exposure outside of burning buildings.
- The rate of firefighter flashover deaths is going up in an environment where the number of fires is going down.
- Otherwise healthy firefighters are experiencing deadly or debilitating heart attacks.

The bottom line: Today's building fire smoke is truly the smoke from hell. Noted Fire Chief, author, and firefighter safety advocate Billy Goldfeder offers this view: "The smoke today is not your Daddy's smoke."

For Discussion:

1. Do you think zero-visibility firefighting fits in today's structural firefighting arena? Why or why not?

2. What changes need to take place in firefighter training to help account for the changes we're experiencing in fire behavior?

3. What types of technological solutions are needed to help firefighters manage volatile smoke?

4. Should ventilation replace rescue as the assumed number one tactical priority at structure fires? Why or why not?

Note:

This case study was inspired by the article, "The Breath from Hell," written by Seattle Fire Officers Mike Gagliano, Casey Phillips, Phil Jose, and Steve Bernocco. The article appeared in the March 2006 issue of *Fire Engineering*.

REVIEW QUESTIONS

1. What is "smoke?"

2. List common hostile fire events and their associated warning signs.

3. What are the four attributes of smoke?

4. How do the four smoke attributes contribute to the understanding of fire behavior within a building?

5. What is meant by the term "Black Fire?"

6. Explain how influencing factors can affect smoke attributes.

7. List the three steps of the reading smoke process.

ADDITIONAL RESOURCES

Gagliano, M., C. Phillips, P. Jose, and S. Bernocco. "The Breath from Hell," *Fire Engineering* (March 2006). Tulsa, OK: PennWell.

Grimwood, Paul. Web site containing great articles and links to fire behavior studies: Available at: www.firetactics.com.

NFPA Fire Protection Handbook, 19th ed. Sec. 3, Chaps. 9 and 10; Sec. 8. Quincy, MA: National Fire Protection Association, 2003.

Quintiere, Dr. James. *Principles of Fire Behavior.* Clifton Park, NY: Delmar Learning, a division of Thomson Learning, 1998.

Wallace, Dr. Deborah. *In the Mouth of the Dragon: Toxic Fires in the Age of Plastics.* Garden City Park, NY: Avery Publishing Group, 1990.

NOTES

1. *Fire Protection Handbook,* 19th ed. Vol. II, Sec. 8 (Quincy, MA: National Fire Protection Association, 2003).

2. James G. Quintiere, *Principles of Fire Behavior* (Clifton Park, NY: Delmar Publishers, a Division of Thomson Learning, 1998).

3. The author has heard Chief Mittendorf say this at seminars, further motivating the development of the reading smoke process included in this text.

4. This is based on research by the author that includes literally hundreds of texts, Internet references, and discussions with fire experts.

5. *Fire Protection Handbook,* 19th ed., Vol. I, Sec. 3 (Quincy, MA: National Fire Protection Association, 2003). The author compared the 2003 *Handbook* fire behavior models to data presented in the 1980s.

6. This is based on the author's research in talking with numerous wood products manufacturers as well as "backyard" testing with components during live-fire training.

Chapter

8

READING RISK

Learning Objectives

Upon completion of this chapter, you should be able to:

- Describe the differences between dangerous and risky.
- List the three influences on risk-taking values.
- List the risk management concepts outlined in NFPA standards.
- Define situational awareness.
- Describe three methods to read risk at an incident.

INTRODUCTION

NFPA 1521 states that the ISO shall monitor conditions to determine if they fall within the department's risk management criteria. How does the ISO monitor conditions to see if they fall within acceptable parameters? What are acceptable parameters? There has been much written and suggested related to risk management, loss control, and risk/benefit thinking. Yet, even with the many resources available, many in the fire service still have a problem figuring out what is acceptable or unacceptable risk-taking at incidents. Saying we have a problem is based on hard research. One of the most glaring proofs that we need work in risk management is firefighter fatality trends. The National Fallen Firefighters Foundation recognized this and held a summit meeting in 2004 to address these trends. The 45 fire service organizations present voted to unanimously endorse 16 Firefighter Life Safety Initiatives. One of the initiatives states:

> Focus greater attention on the integration of risk management with incident management at all levels, including strategic, tactical, and planning responsibilities.[1]

At an incident, the ISO must read the risks being taken and offer judgment on their acceptability. This chapter looks at firefighter risk taking and offers thoughts to help the ISO read risk and offer sound judgments regarding its acceptability.

FIREFIGHTER RISK TAKING

■ **Note**
Firefighting isn't dangerous, it's merely risky. This perspective came from Chief Dave Daniels of the Renton Fire Department (Washington) in addressing participants at a recent Incident Safety Officer Academy.

Historically, firefighters have endeared themselves to the communities they serve through selfless acts of bravery and sacrifice, and in fact most people believe that the firefighting profession is dangerous. Ask any firefighter if firefighting is dangerous, and the response may be a humble "sometimes" or a resounding "hell, yes!" Now consider a different perspective: "Firefighting isn't dangerous, it's merely risky." This perspective came from Chief Dave Daniels of the Renton Fire Department (Washington) in addressing participants at a recent Incident Safety Officer Academy. The point is eloquent and should be adopted as an "attitude" by the ISO. Look at it this way. From a community perspective, firefighters are expected to work in inherently dangerous environments. From a fire service perspective, the risks of many specific dangers are well-known. Further, we learn, train, and equip ourselves to understand the dangers and take steps to avoid, control, or eliminate the dangers. When we view things this way, we make *choices* about the dangers we face. That is risk taking **(Figure 8-1)**.

Figure 8-1 *Fire-fighters make choices about the dangers they face; that is risk taking. (Photo by Keith Muratori from FIREGROUND-IMAGES.com.)*

Firefighters are action oriented and results oriented: What a wonderful problem to have! However, this action and results orientation is also the occasion for injuries and deaths because many firefighters are arbitrarily aggressive; their action approach to get results exposes them to dangers that they may not appreciate. Since the causes of most firefighter fatalities are well-known—along with the situations that cause them—it seems reasonable that we can make decisions to avoid or control those situations. Instead of being *arbitrarily* aggressive, we should be *intellectually* aggressive. Being intellectually aggressive requires being able to *recognize* predictable dangers and taking steps to reduce risk taking by our fellow responders. You can acquire this perspective by front-loading two things: the understanding of defined risk-taking values and the increased ability to achieve situational awareness.

> **Safety**
> Instead of being *arbitrarily* aggressive, we should be *intellectually* aggressive.

RISK-TAKING VALUES

Determining the appropriateness of risk taking is perhaps the most difficult decision that the ISO has to make at the incident scene. In most cases, the incident commander has already established risk boundaries for working crews. For example, the IC who has pulled interior crews and deployed exterior defensive fire streams has, in essence, determined that firefighters may no longer risk interior firefighting. In other cases, the line is not so clear. This is where the ISO has to make the *value* decision of whether a specific task, strategy, or action is worth the risk of injury. Some have expressed this as deciding risk/benefit. As an ISO making value decisions, you must first understand what values have been defined by the factors discussed in the following sections.

Community Expectations

Most firefighters accept the notion that they may have to risk their lives to save a life. In most cases, the community served by the firefighter expects that of the firefighter and the fire department. This basic expectation is what separates the firefighting profession from many others and gives it a status of respect in the community.

Firefighters who swear to uphold the values of the community (to protect life and property) must draw on courage and in some cases bravery to meet the community's expectations. The same courage and bravery, however, must be tempered with a heavy dose of prudent judgment so that a situation does not unnecessarily harm the firefighter. The key word is "unnecessarily." When a firefighter dies protecting property (as opposed to life), the taxpayer may question whether prudent judgment prevailed. Clearly, the community does not expect firefighters to die needlessly. This is where the IC and the ISO must step in and establish a pace, an approach, or a set of guidelines to avoid an unnecessary injury.

Advances in media communication have increased the community expectations of firefighters. The media loves to depict firefighters taking high risks at fires, unusual rescues, and disasters—it makes good copy. When John and Jane Doe watch these events, they may mistakenly believe that their own firefighters are trained and ready to perform in the same way for their community. This puts risk-taking pressure on the local fire service that is not equipped or trained for such an event—yet the community expects that kind of service because they saw other firefighters doing it on the news. In many ways, the fire service has become the agency of "first and last resort" for events that affect the public. Again, this puts risk-taking pressure on responders.

Fire Service Standards

NFPA

NFPA standards specifically address risk management concepts related to the handling of emergency operations. The concepts are repeated in NFPA 1500, 1561, and 1521.

NFPA standards specifically address risk management concepts related to the handling of emergency operations. The concepts are repeated in NFPA 1500, 1561, and 1521. These concepts are spelled out quite clearly and should be incorporated into the incident action plan. From these concepts, risk-taking decisions can be made (see Risk Management Principles [NFPA 1500]).

Risk Management Principles (NFPA 1500)

1. Activities that present a significant risk to the safety of members shall be limited to situations in which there is a potential to save endangered lives.

2. Activities routinely employed to protect property shall be recognized as inherent risks to the safety of members, and actions shall be taken to reduce or avoid those risks.

(continued)

> (*continued*)
> 3. No risk to the safety of members shall be acceptable when there is no possibility to save lives or property.
> 4. In situations where the risk to fire department members is excessive, activities shall be limited to defensive operations.[2]

Most fire officers would subscribe to the risk principles in the NFPA standards. Simply stated, these concepts can be shortened and made easy to remember with the following brief statements:

Risk a life to save a known life.

Perform in a predictable, practiced manner to save valued property.

Take no risk to save what's lost.

Default to defensive when conditions deteriorate quickly.

These statements may seem black and white to some, but experience tells us that their interpretation is still wide-open. ISOs might have to be more discreet than others. For example, the first statement says, "Risk a life to save a known life." It does not say "sacrifice" a life to save a known life. Throwing firefighters at a rescue may not be the answer. Likewise, the statement is not plural. At some point, the IC or ISO must determine how many firefighters should be at risk to save a life.

valued property
physical property whose loss will cause harm to the community

When deciding on a "practiced, predictable manner" to save valued property, the ISO should first determine what "valued property" is. While definitions may vary, one criterion seems clear. **Valued property** can best be defined as physical property whose loss will cause harm to the community. Examples are a hospital, a significant employer, a utility infrastructure, or a place with historical significance. In these cases, the ISO should then calculate the risk (the possibility and severity of an injury), then weigh safety factors in favor of the firefighter. In other words, do what is practiced and routine. For example, an interior firefight at a county clerk's office may be extended because of the potential for the loss of irreplaceable records. The ISO may recommend that additional engine companies be ordered for backup or that emergency roof and ceiling bracing be installed to prolong the time the firefighters can fight or protect salvage operations.

Department Values and Skills

When determining an acceptable or unacceptable risk, ISOs must consider what is commonplace and accepted by their departments in terms of acceptable risks. In one department, the aggressive attack on an interior fire may not be acceptable until ventilation efforts are underway. Yet, another department may decide that it is acceptable to start a fire attack before ventilation is accomplished. The first department places an extreme value on ventilation. Conversely, a small, rural Kansas Fire Department, may decide that *no* interior fire attack is allowed on anything other than an incipient fire; the department simply does not have the equipment and training for that level of risk.

The ISO must evaluate an operation and decide whether the situation fits the organization's "normal" way of handling the incident. Such a decision can easily lead to conflict. The department may have embraced an inherently unsafe way of doing a task. A good example is the department that allows freelancing due to insufficient on-scene staffing or the department that routinely launches an interior fire attack before four people have assembled on scene. In these cases, the ISO must be extra alert and should try to build a bridge toward a safer operation. These cases try the ISO and present a challenge to reduce injury potential creatively.

Regarding skills, the ISO should recognize when crews are attempting to perform a skill for which they have never prepared. The rise-to-the-task attitude of most firefighters creates such situations. In these cases, the ISO should encourage technical assistance, a slower pace, and an increase of zoning or backup.

As you can see, the foundation for risk taking is established by defined values. To apply these values, you must be keenly dialed into the incident. We call this situational awareness.

SITUATIONAL AWARENESS

In a simpler time, the term "size-up" described the process used to become aware of the incident situation. In the 1970s, we were exposed to a simple five-step size-up. In the eighties it was 13-point size-up. In the nineties it was 27-point . . . you get the picture. The complexities of incident handling have increased exponentially, and fire service leaders attempt to create and define size-up models to help us mentally process an incident. These efforts are to be commended.

Despite such efforts, the ISO needs a method to rapidly "get" what is happening at an incident and the risks at play. Instead of using a size-up model, consider using a situational awareness approach that helps you "read risk." Situational awareness, in this context, is different from what fire officers are usually thinking about. (If you were to ask fire officers at an incident if they are aware of the situation, they would probably give you a curious look and the knowing, "Excuse me?") Situational awareness is actually a measurement of accuracy. In fact, the Naval Aviation Schools Command in Pensacola, Florida, defines **situational awareness** as the degree of

situational awareness
the degree of accuracy by which one's perception of the current environment mirrors reality; applied to the ISO, situational awareness is the ability to accurately read potential risks and recognize factors that influence the incident outcome

accuracy by which one's perception of his current environment mirrors reality. Similarly, ISOs' situational awareness is their ability to accurately read potential risks and recognize factors that influence the incident outcome. There are many ways to achieve accuracy in awareness. To improve your ability to achieve situational awareness, you must be keenly aware of the factors that reduce your ability and strive to keep your mind open. Keeping your mind open is easy to say—and hard to achieve. Seeking out information, dropping biases, and striving for accuracy are all ways to keep your mind open. (To see what does not help, see Factors that Reduce Situational Awareness.)

Factors that Reduce Situational Awareness*

- Insufficient communication
- Fatigue and stress
- Task overload
- Task underload
- Group mind-set and biases
- "Press on regardless" philosophy
- Degrading operating conditions

---

*Source: Naval Aviation Schools Command, Pensacola, Florida.

To help you apply situation awareness to reading risk at an incident, let's look at several methods that have been developed to help you "get" the risks being taken on.

The Brunacini Approach

Former Chief Alan Brunacini and his son, Battalion Chief Nick Brunacini of the Phoenix Fire Department, coauthored a book called *Command Safety* in 2004.[3] In their text, the authors took the concept of situational evaluation (Chapter 2 of their book) and defined a series of "gauges" to help an IC understand hazard severity. Each gauge included a green-yellow-red scale of relative dangers to responders. They measured many factors that have historically led to firefighter fatalities and used them to help the IC realize the potential for a negative outcome. While designed for the IC, the gauges are applicable to the ISO role. As you can see in The Brunacinis' Situational Evaluation Factors text box, the chiefs from Phoenix biased the factors toward structural fires. With a little imagination, you can apply the approach to a multitude of incident types.

The Brunacinis' Situational Evaluation Factors

Each of the factors is "gauged" 1–5 with 5 being the highest risk.*

1. Overall risk level
2. Building size/area
3. Fire stage
4. Penetration distance
5. Heat level
6. Percentage of involvement
7. Smoke conditions
8. Structural stability
9. Fire load
10. Occupancy hazard
11. Residential/commercial relativity
12. Access/exit issues
13. Interior arrangement
14. Aggressiveness
15. IC's instinct
16. Red flags (a list of "historic losers")

*Some of these factors are paraphrased by the author.

Value—Time—Size Method

Stewart Rose, retired safety chief from the Seattle Fire Department, designed a class entitled "Managing Multiple Company Tactical Operations." In this class, Chief Rose presented a quick risk-versus-benefit evaluation that offered a way to think through the appropriateness of risk taking. Simply stated, you evaluate value, time, and size (see Stewart Rose's Risk-Versus-Benefit Evaluation).

Stewart Rose's Risk-Versus-Benefit Evaluation

To determine risk versus benefit, determine the following (value—time—size thinking):

1. Can something be saved (the value)?
   People
   Buildings

2. What is a safe time for firefighting, based on construction and the location of the fire (the time window)?

3. What is the amount of water needed to extinguish the fire (the size)?
   Hand line(s)
   Master stream

Again, with a little imagination, you can take the value-time-size concept and apply it to numerous types of incidents. For example, in the case of a mud slide with multiple structural collapses, the value remains unchanged (people or property), time is the window of opportunity you have compared to the stability of the mud and structures, and size applies to the amount of resources that need to be deployed to affect mitigation.

The ISO's Read-Risk Approach

■ **Note**

Knowledge, sound judgment, experience, and wisdom are paramount in making risk decisions.

The two preceding methods of reading risk are well accepted in the fire service and can serve the ISO effectively. Over time, experienced ISOs have developed their own approach to reading risk **(Figure 8-2)**. You too can draw from many sources to determine whether a situation presents an acceptable risk. Knowledge, sound judgment, experience, and wisdom are paramount in making risk decisions. One way to prepare for situational risk decision making is to read the many accident investigation reports generated for firefighter duty-deaths. This is vicarious learning and can help prevent yet another firefighter death. **Vicarious learning** can be defined as learning from the mistakes of others.

vicarious learning
learning from the mistakes of others

As an example, based on my own reading and research, I have developed a "read-risk" approach that might be useful in your role as an ISO. The approach consists of a collection of the information and questions, as presented in this section.

> *Step 1: Collect information:*
> - Read the building.
> - Read the smoke.

Figure 8-2
Experienced ISOs typically develop their own process for reading risk at incidents.

- Read firefighter effectiveness. Are we doing predictable things and getting standard outcomes?

Step 2: Analyze:

- Define the principal hazard. (What is going to hurt the responders?)
- What is the window of opportunity (time)?
- Are we ahead or behind the power curve?
- What is really to be gained?

Step 3: Judge risk:

- Are we within the risk-taking values established by the department?
- Are we doing all we can to continually reduce risks?

In many ways, these questions force the ISO to think ahead and predict hazard potentials. This is proactive thinking—the hallmark of an effective ISO.

Regardless of the method you use to read risk, follow through and make a judgment regarding risk taking. If the judgment you make does not fit the incident plan, action is warranted. If you are uncomfortable with the risks being taken but cannot quite justify or articulate your concern, then that, in itself, is reason to visit with the incident commander. You may find it effective to approach the IC and ask how he or she feels about the situation—sort of a "reality check." This is an example of being a consultant and a helper in the team approach to managing risks, which we talk about more in Section Three.

SUMMARY

At an incident, the ISO must read the risks being taken and make a judgment on their acceptability. Firefighters work in dangerous environments. Yet the dangers are mostly known, and we take many precautions to limit the dangerous exposure. This makes firefighting risky, not necessarily dangerous. Over time, firefighters have become arbitrarily aggressive as opposed to intellectually aggressive. Judging acceptable and unacceptable risks starts with an understanding of the risk-taking values in play, which are defined by community expectations, fire service standards, and fire department values and skills. At an incident, the ISO must employ situation awareness techniques to help evaluate risks. Although many size-up techniques exist, the ISO should utilize a method that is focused on risk taking. Three methods can achieve this: the Brunacini approach, value-time-size thinking, or the ISO's read-risk approach.

KEY TERMS

Situational awareness The degree of accuracy by which one's perception of the current environment mirrors reality. Applied to the ISO, situational awareness is the ability to accurately read potential risks and recognize factors that influence the incident outcome.

Valued property Physical property whose loss will cause harm to the community.

Vicarious learning Learning from the mistakes of others.

REVIEW QUESTIONS

1. Describe the differences between Dangerous and Risky.
2. List the three influences on risk-taking values.
3. List the risk management concepts outlined in NFPA standards.
4. What is "valued property"?
5. What is meant by situational awareness?
6. Describe three methods to read risk at an incident.
7. What is vicarious learning?

ADDITIONAL RESOURCES

Emery, Mark. "13 Incident Indiscretions." *Health & Safety for the Emergency Service Personnel* 15, no. 10 (October 2004).

Federal Emergency Management Agency, United States Fire Administration. *Risk Management Practices in the Fire Service,* FA-166 (December 1996). Available at: www.usfa. fema.gov/application/publications.

United States Fire Administration. *Firefighter Life Safety Initiatives.* Emmitsburg, MD: National Firefighters Foundation, 2004. Available at: www.everyonegoeshome.com.

NOTES

1. United States Fire Administration, *Firefighter Life Safety Initiatives* (Emmitsburg, MD: National Firefighters Foundation, 2004).
2. NFPA 1500, *Fire Department Occupational Safety and Health Program,* 8.3.2 (Quincy, MA: National Fire Protection Association, 2007).
3. A. Brunacini and N. Brunacini, *Command Safety* (Phoenix, AZ: Across the Street Productions, 2004).

Chapter 9

READING HAZARDOUS ENERGY

Learning Objectives

Upon completion of this chapter, you should be able to:

- Define hazardous energy and list four ways to categorize its status.
- List common electrical equipment and their associated hazards.
- List the chemical properties of common utility gases.
- List the hazards associated with utility water and storm sewer systems.
- Give examples of mechanical hazardous energy.
- List the hazardous energy sources in vehicles.
- Discuss weather as hazardous energy and itemize the warning signs that extreme weather is approaching.
- Describe why water is a form of hazardous energy.

DEFINING HAZARDOUS ENERGY

Ask a firefighter to define hazardous energy and you will likely hear a list containing physically hazardous energy sources like electricity and pressure vessels. While these are certainly *forms* of hazardous energy, they do not serve well as definitions. In the context of this book, **hazardous energy** is defined as stored potential energy that will cause harm if suddenly released.

The modern, technologically advanced society is full of hazardous energy sources—to the point that most of it is taken for granted. Although we have engineered safety into most hazardous energy sources, this sometimes only sets the stage for complacency. Emergency responders see the results of this complacency often—and unfortunately—as we rescue victims. Most firefighters appreciate and respect the dangers associated with electrical equipment, pressure vessels, and hazardous chemicals. Still, the demands of an incident may place firefighters in harm's way should one of these hazardous energy forms be released. This is where the ISO comes in.

To be an effective and efficient ISO, you must front-load your understanding of hazardous energy forms so that you can better predict its release and intervene appropriately to protect fellow responders. You must analyze each form of hazardous energy and determine its degree of potential impact on the incident. If you understand the hazards associated with a hazardous energy source, you can compare what you know to the incident conditions and make a judgment about the stability of the source. Usually we do this by categorizing the energy as stable or unstable. The effective ISO categorizes the status of hazardous energy forms in one of the following at incidents:

- Stable—not likely to change
- Stable—may change
- Unstable—may require attention
- Unstable—requires immediate attention

This chapter lists hazardous energy forms and furnishes information on them so that you can better understand their hazardous potential.

hazardous energy
stored potential energy that causes harm if suddenly released

FORMS OF HAZARDOUS ENERGY

Electricity

The integrity of electrical systems is based on their being properly grounded, insulated, and circuit protected. A disruption of any of these components can pose a danger to firefighters. **Figure 9-1** outlines basic electrical terms. The ISO who is familiar with them can better communicate electrical hazards and concerns.

Electrical Terms

Ampere (Amps)
The unit of measure for VOLUME of current flow.

Continuity
The completeness of a current circuit; the ability of electricity to pass unbroken from one point to the next.

Conductivity
The measure of the ability of a material to pass electrical current.

Energized
The presence of electrical current within a material or component.

Ground
The position or portion of an electrical circuit that is at zero potential with respect to earth. The point of electrical return for an electrical circuit.

Ground Fault Interruption (GFI)
A device that will break continuity in a circuit when grounding occurs before the current returns to the distribution source.

Grounding
The act or event of creating a point for electrical current to return to zero potential.

Ohm
The unit of measure for resistance of electrical current. One ohm equals the resistance of 1 volt across a terminal at 1 ampere.

Resistance
The degree that a material holds or impedes the flow of electrical current.

Static Discharge
The release of electrical energy that has accumulated on an insulated body. Static is a stationary charge looking for a ground.

Voltage/Volt
The FORCE that causes the flow of electricity. A volt is the unit of measure for electrical force or potential.

Watts
A unit of measure for the amount of energy a specific appliance uses.

Figure 9-1 *The ISO should be comfortable with electrical terms and be able to communicate hazards accurately.*

Fire and rescue departments are often called to incidents that involve electrical distribution equipment (cat on the pole, wires down, and the like), but rarely do firefighters have to handle the incident without the assistance of the local power company. Occasionally, however, firefighters must act without the power company to save lives or stop a potential loss. In these cases, a basic understanding of electric system integrity is important. For the incident safety officer, the understanding of these systems is *essential*.

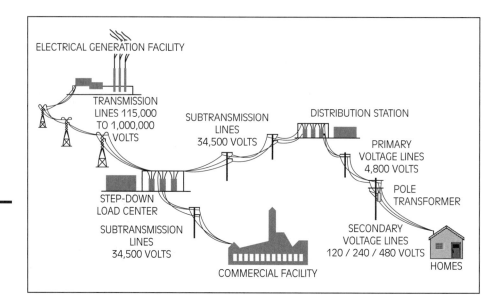

ELECTRICAL GENERATION FACILITY

TRANSMISSION LINES 115,000 TO 1,000,000 VOLTS

SUBTRANSMISSION LINES 34,500 VOLTS

DISTRIBUTION STATION

PRIMARY VOLTAGE LINES 4,800 VOLTS

STEP-DOWN LOAD CENTER

SUBTRANSMISSION LINES 34,500 VOLTS

POLE TRANSFORMER

SECONDARY VOLTAGE LINES 120 / 240 / 480 VOLTS

HOMES

COMMERCIAL FACILITY

Figure 9-2 *ISOs must be able to recognize the components of a typical municipal electrical grid.*

■ Note

Ground gradient is electrical energy that has established a path to ground through the earth and is energizing it. A downed power line may be energizing the earth in a concentric ring of up to 30 feet depending on the voltage of the source.

ground gradient
electrical energy that has established a path to ground through the earth and that continues to energize the earth; an example is a downed power line

The best way to learn about them is to contact your local power company; most provide no-cost seminars and workshops for firefighters. In the scope of this book, we concentrate on a few *recognition* concepts for the ISO. **Figure 9-2** shows a typical electrical generation and distribution system. Along the chain of distribution, various equipment components handle the electricity. **Figure 9-3** lists some of the specific hazards associated with this equipment. Being familiar with the arrangement of overhead equipment, cables, and power lines on power poles may help the ISO recognize potential hazards. **Figure 9-4** shows a typical overhead power distribution setup, although it is important to understand that there are numerous configurations of overhead power equipment.

Electricity is always trying to seek the path of least resistance to ground. Incidents involving electrical equipment, such as pad transformers and downed wires, are especially dangerous. In these cases, electricity may be seeking ground through vehicles, fences, and other conductors. A victim in an energized car is like a bird on an overhead power wire, that is, as long as the ground is not touched, no electrocution can take place. Anything or anyone who touches the vehicle and the ground at the same time can create a path for electricity to seek ground. Firefighters attempting to approach compromised electrical equipment may feel a tingling-sensation in their boots; this is a danger sign known as ground gradient. **Ground gradient** is electrical energy that has established a path to ground through the earth and is energizing it. A downed power line may be energizing the earth in a concentric ring of up to 30 feet depending on the voltage of the source. A firefighter who feels

Electrical Equipment Hazards

Powerlines/Wires

Uninsulated, under tension, arc danger, difficult to know voltage/amperage, downed wires may jump/recoil, ground/fences/gutters easily energized when in contact, power feed may be from both directions.

Pole-Mounted Transformers

Usually step-down type, difficult to extinguish, may drop/dangle, may cause pole damage, may cause wire failure, may drip hot oil and start ground fire.

Pad-Mounted Transformers

Usually low-voltage/high-amperage, may energize surrounding surfaces, pooled water may conduct current, difficult to extinguish oil/pitch fire, possibility of arcing.

Ground Level Vaults

Confined space, possible O_2 deficiency, buildup of explosive gases/smoke, cable tunnels can transfer fire/heat, significant arc danger.

Subterranean Vaults

Same as ground level vaults, plus water collection hazard, difficult to ventilate, can "launch" a manhole cover if accumulated gases ignite.

Generators

Power source, tremendous heat generation, hazards of diesel/gasoline/gas-fueled engines, automatic startup when other power sources are disconnected.

Batteries

Stored energy, chemical/spill hazards, explosive gas buildup, multiple exposed terminals.

Disconnects/Switches/Meters

Exposed terminals, danger of arcing.

Figure 9-3 *Electrical components can present hazards to firefighters.*

> **Safety**
> At all incidents, the ISO should evaluate the proximity and integrity of electrical systems. When a component of the total system is deemed to have lost integrity, the ISO must assume that an electrical danger exists and communicate that assumption to responders.

tingling in his or her boots must back away using a shuffle-foot motion to keep both feet in contact with the ground. In all cases, the fire department must verify that power company technicians have been dispatched. Some ISOs carry specialized electrical-current–sensing equipment like a "hotstick" or "voltage alerting pens" to detect electrical current movement through objects. As with any specialized tool, training and understanding of the equipment's limitations are essential.

At all incidents, the ISO should evaluate the proximity and integrity of electrical systems. When a component of the total system is deemed to have lost integrity, the ISO must assume that an electrical danger exists and communicate that assumption to responders. In some cases, the electrical system can be perfectly intact and still pose a threat to firefighters, particularly when aerial apparatus and aluminum ladders are being used near power lines and

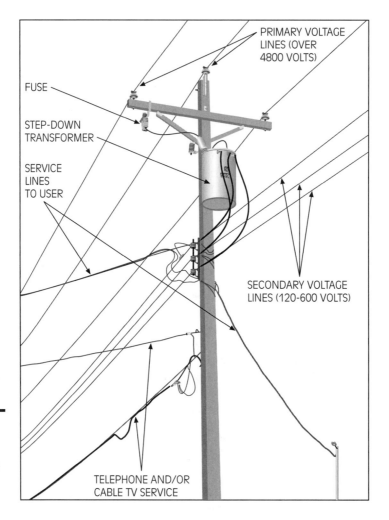

Figure 9-4 *This typical overhead power pole arrangement is just one of the many possible.*

electrical equipment. Although it happens rarely, power can "jump" conductors if a conductor comes close to the exposed wire or power connection. **Table 9-1** shows the 360-degree minimum operating distances for equipment operating near electrical lines.

Incident operations involving electrical equipment are numerous, and specialized tactics should not be attempted without consultation with the electrical power authority. In a few specific electrical environments, the ISO (and responders) must exercise particular attention:

- *Battery rooms:* The increased reliance on computers and data transmission systems has led to an increase in the presence, and capacity

TABLE 9-1   *Minimum distance requirements when working around electrical equipment.*

Voltage	Distance (feet)
0–50,000	10
50,000 to 200,000	15
200,000 to 500,000	20
500,000 to 700,000	32

of, uninterrupted power supplies (UPS). UPS battery rooms are commonplace in many data storage and processing centers, as well as in telecommunication switch equipment centers. In such installations, it is not uncommon to use reconditioned wet cell batteries from submarines for UPS purposes, and some of these battery banks can take up more than several thousand square feet of floor space. Emergency responders working in or around UPS rooms need to exercise extreme caution because of the presence of direct current (DC) power for the batteries. Pooled water, battery acid, and battery rack hardware can become DC energized because there is typically no ground fault protection with DC power.

- *Substation fires:* Electrical transmission and distribution substations use large transformers to step down voltage. Incidents involving this equipment present several hazards to responders. These transformers can hold several thousand gallons of oil. When a transformer catches fire, it may burn for several days, and the heat generated from an oil-fed fire can cause serious damage to the high-tension wire infrastructure and other transformers. In most cases, the power company makes the decisions to have the steel/aluminum infrastructure cooled and the fire to be extinguished. Substations are typically well drained so water cannot pool, but make sure the drainage systems are working. High tension wires that separate from anchor points can "reel coil" with tremendous force; that is, when the tension is released suddenly, the wire snaps, or whips, in a direction opposite to the tension force. When that happens, two energies are at play. First is the law of motion: For every action (or force), there is an opposite and equal action (force). Second, steel retains a certain memory. Most steel cable is tightly wound onto a spool during production. When it is released from its application, it returns to its reeled state; hence "reel coil." Make sure responders stay clear of "whip" pathways.

Utility Gas

Firefighters who have seen a propane tank BLEVE (boiling liquid expanding vapor explosion) undoubtedly agree that the most important utility to control is gas. The ISO should evaluate the integrity of the gas fuel supply and containment vessels. The integrity of a gas system relies on a tight supply vessel (a tank or piping), a shut off valve, a pressure regulation device, and a distribution system with protection at each appliance (shutoff valve and surge protection). Each component in the system is designed for a certain pressure. Pressures that exceed the maximum (such as from exposure to heat from a fire) can cause a pressure relief device to activate, introducing expanding gas into the environment. Likewise, trauma to this system can create holes, pipe separation, or container failure. Once gas escapes from the system, fires may be accelerated or toxins can be released and inhaled. Unignited gases can accumulate in confined spaces and present an explosion hazard if an ignition source and proper air mixtures are introduced. Knowing the properties of common gases can help the ISO determine the risks associated with the gas present **(Figure 9-5)**.

Utility Water and Storm Sewer Systems

Some may argue that utility water and sewer systems are not hazardous energy. Anyone who has witnessed a rapidly enlarging sink hole after a waterline break readily asserts that that water is indeed hazardous energy. Water and storm sewer systems can cause safety concerns in numerous incident situations. Uncontrolled water flow can cause initial and secondary collapses in structures. A damaged sewer or storm drain system can leech

Figure 9-5 *The ISO who knows gas properties can more effectively assess risks associated with utilities.*

Properties of Common Utility Gases		
Property	**Propane (Liquefied Petroleum Gas)**	**Natural Gas (Methane)**
Chemical makeup	C_3H_8	CH_4
Vapor density	1.5	.55
Boiling point	−44°F	−259°F
Ignition temperature	871°F	999°F
Flammability limits		
Upper	9.5	14.0
Lower	2.4	5.3

water into surrounding gravel and dirt beds, undermining the earth supporting a structure, road, or other utility system. Firefighting efforts may introduce significant quantities of water into a structure. Who evaluates where all that water is going to go? The ISO should investigate large quantities of water flowing in and around the incident. In one incident, the assigned ISO saw large quantities of water pooling and bubbling in the yard next to a dwelling fire, which seemed unusual. Upon investigation, the ISO determined that the piping feeding the hydrant being used for water supply had begun leaking underground. The pooling was the first visual clue that the ground underneath the firefighting operation was being undermined. The water department was notified, and the water shut down but not before nearby street pavement began sinking next to a staged engine.

Water that collects in basements or other building areas may extinguish gas-fired equipment pilot lights, with the result that raw gas is being "bubbled" into the water. Although many gas-fired devices have fail-safe systems to stop raw gas from flowing, in a fire, after a collapse, or in waterlogged conditions, these fail-safes may be damaged. Likewise, pooled water and electrical equipment do not usually play well together.

Storm sewer systems are designed to receive potentially large volumes of water, literally thousands of gallons per minute! In flash flood situations, these systems can be overwhelmed with water and debris. The debris can get caught up at storm grates and cause localized flooding. The tremendous force of the flowing water can pull unsuspecting victims into the debris and grates, pinning them below water levels or crushing them with force. Firefighters have been killed attempting to rescue victims caught in storm sewer systems. The use of swift water rescue techniques and solid risk management concepts need to be employed when working around storm sewer systems during flash floods.

Mechanical Energy

Systems that include pulleys, cables, counterweights, and springs are examples of mechanical stored energy. Lightweight high-rack storage systems can be classified as stored potential energy in the form of dead or live loads and gravity. The sudden release of mechanical systems can be caused by heat, trauma, and/or overloading. Steel cables used in pulley systems are likely to recoil violently when they fail. Simple "guide," or "guy," wires for poles, signage, and antennae are typically tensioned and can release and recoil with amazing force. Freestanding truss structures, like antennae, are nothing more than vertical cantilevered beams and are weakened quickly when exposed to the heat of a fire. These weakened towers fail quickly, and horizontal forces like wind or blunt force accelerate the collapse.

▮ Safety

● Simple "guide," or "guy," wires for poles, signage, and antennae are typically tensioned and can release and recoil with amazing force.

Pressurized Systems and Vessels

Pressurized equipment comes in many forms, sizes, mediums, and arrangements. Regardless, suddenly released pressure from an enclosed container can literally slice through a person. Pressurized systems use either hydraulics (liquids) or pneumatics (air/gases) as a medium to achieve power or force. Pressurized systems typically include a reservoir for the medium, some sort of pump, a tubing or conveyance system, and the tool or implement that uses the pressure to achieve a task. Connections, valves, manifolds, and pressure control devices may also be present. Heat applied to a pressurized system causes pressure to increase, which can exceed the system's design limits. Built-in pressure relief devices may not be able to relieve pressure as fast as the pressure is being developed, leading to an explosive failure of a component.

Closed containers of various products (mainly liquids and some gases) may become "pressure vessels" when heated by fire or hot smoke. While gaseous product containers are designed to hold certain pressures, the fire can cause the pressure to exceed design limits. Failure of the container can cause the container to become an unguided missile. Liquid containers may or may not be designed to hold pressure, yet pressure is developed within them when they are heated. Again, they may explode or become missiles. As a rule, the stronger the container is, the more initial resistance to pressure it has and the more explosive it becomes when it fails.

Hazardous Energy in Vehicles

In this text, all sorts of transportation equipment is classified as vehicles. Cars, trucks, boats, planes, trains, and the like are included in the classification. Vehicles contain many forms of hazardous energy. Their systems and hazards may include:

- *Stability/position:* Rolling weight, instability, collapse (crush), failure of ground to hold the load
- *Fuel systems:* Types of fuels, storage, pumps, fuel lines, pressurization
- *Electrical systems:* Batteries, converters, high-voltage wires
- *Power generation systems:* Pulleys, belts, heat, noise, thrust, exhaust gases
- *Suspension/door systems:* Springs, shocks, gas or pneumatic struts
- *Drive/brake systems:* Pressure vessels, heat, springs, torsion, exotic metal fumes
- *Restraint safety systems (air bags):* Trigger systems, chemicals, delayed or unpredicted deployment

The list can go on, but a few specific areas need to be addressed. The accidental or delayed deployment of air restraint devices can cause—and has caused—injury to rescuers. C-spine trauma, muscle sprain, contusions, and lacerations are all possible if the unsuspecting firefighter is suddenly hit by a late deploying restraint bag while attempting to extricate victims of vehicle collisions. A victim trapped in a vehicle with an undeployed air restraint safety device is in danger; consider this an immediate rescue environment. In some cases, the actions of firefighters can cause the activation of an air restraint device. While it is beyond the scope of this book to discuss all the steps to deactivate air restraint devices, responders should research and develop procedures for restraint device deactivation.

New fuel and power system technologies are being developed to address traditional fuel supply shortages and dependency on foreign oil. Alternative fuels, fuel cells, and high-voltage systems present additional hazards to responders. Many of our educational resources (books, suggested procedures, and incident experience) trail the incident potential of these hybrid systems. Identifying a vehicle with an "alternative" or hybrid power system can be challenging. A few "standards" are in place but they are likely to change as manufacturers develop new and improved systems. Although the standard is subject to change, bright orange conduit or cable can indicate that high-voltage power (up to 700 volts and 125 amps) is used in the drive train of a vehicle. These high-voltage systems typically have large battery packs to help store and feed energy. Bright blue conduits or cables can indicate medium-voltage drive trains (less than high-voltage systems but more than a typical 12- or 24-volt systems). Ethanol fuels (like E-85) are polar solvents that can render class B foam ineffective; use alcohol-resistive foams for spill and fire control. Err on the side of caution when dealing with these "advances."

Weather

"If you don't like the weather, just wait an hour and it will change!" This saying is especially true in many regions of the United States. In fact, weather is a dynamic, complex, and often misunderstood force that firefighters must contend with. Common fire house talk suggests that the "big one" will hit when the weather is at its worst (perhaps just another example of Murphy's law). Often, the effects of adverse weather cause the reporting party to call for firefighting service. Progressive fire and rescue departments have acknowledged this possibility and have taken steps to reduce the impact of adverse weather on personnel. Once on-scene, the ISO should consider weather as a form of hazardous energy and weigh the effects of weather extremes with the "behavior" of the incident and of the incident responders.

Effective ISOs study weather and understand the particulars of weather patterns found in their geographical regions. Likewise, they keep abreast of daily forecasts and weather observations as a matter of habit and readiness. From this constant attentiveness, they can make reasonable predictions on the impact of weather-related risks.

In-depth professional weather spotter and forecasting training is available from local television weather forecasters and the National Weather Service (often free of charge). The National Weather Service is part of the U.S. Commerce Department's National Oceanic and Atmospheric Administration (NOAA), which has an extensive collection of pamphlets, books, videos, and charts available for weather-related education. This section of the book only highlights weather considerations essential to effective ISO performance. These considerations are wind, humidity, temperature, and the potential for change.

Wind   Of all the weather considerations affecting firefighting operations, wind is by far the most important. Nothing can change a situation faster—or cause more frustration in an operation—than the effects of wind. Wind is created as air masses attempt to reach equilibrium. Warm and cold fronts cause changes in atmospheric pressure and therefore gradients in pressure. Other factors, such as the jet stream (the prevailing wind), day/night effect (the diurnal wind), upslope/downslope, and sea breeze, all influence wind. The arrival of a cold front can cause a 180-degree change in the wind direction. A falling barometer (the measure of atmospheric pressure) can indicate an approaching storm and subsequent wind. A warm, dry wind that spills down through mountain canyons or valleys can be extremely dangerous to firefighters in that it quickly changes fire behavior (whether wildland or structural). Likewise, the sudden reduction of a warm wind allows the local or prevailing wind to influence the fire. A sudden calm period may indicate an upcoming weather event, such as a thunderstorm downdraft or a wind shift.

Some indicators of wind and wind changes can be found in the patterns formed by clouds. High, fast moving clouds may indicate a coming change, especially if the clouds are moving in a different direction than the surface wind. Lenticular clouds (very light, sail-shaped cloud formations found high aloft) indicate high winds that may produce strong downslope winds if they surface. The development and subsequent release of a thunderstorm causes erratic winds and strong downdrafts (**Figure 9-6**). It is not uncommon to have four or five wind direction and speed changes in the vicinity of a thunderstorm squall line.

ISOs should understand the wind patterns for their specific geographical region. Further, the ISO should evaluate existing and forecasted winds and predict their influence on fire behavior and the safety of responders. As wind velocity increases, so does the risk to firefighters (see High Wind Danger at Structural Fires). Knowing local wind tendencies and evaluating their effects on an incident can make a difference between a high-risk and a relatively safe operation!

High Wind Danger at Structural Fires

Firefighters are well aware of the impact high winds can have on outside or wildland fires, but their impact on structural fires is equally important. Of particular
note are the choices that firefighters make for the point of fire attack and the point
of ventilation. One age-old tenet of fire stream application is to attack from the
unburned side. While this principle is appropriate sometimes, we know that at
times it is not possible, practical, or desirable. High wind influence should be
considered when making attack point decisions. Open doors, sudden window
failure, and other vent spaces can allow wind to penetrate the interior space of the
building and influence fire behavior. All firefighters should be aware of this
possibility, and the ISO should evaluate attack options to see if firefighters are in
a position to be trapped, overrun, or exposed to a wind-fed acceleration of the fire
inside the structure. Likewise, the influence of wind should be part of the decision-
making process when locating ventilation openings. Similarly, a strong wind can
easily defeat the desired outcome of PPV fan use.

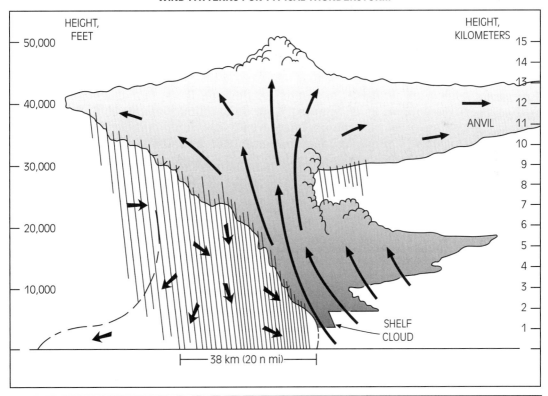

Figure 9-6 *Erratic wind shifts are common during thunderstorms. The ISO needs to watch weather*
influences during incidents.

Humidity   The ISO working a large wildland fire has likely established communication with the fire behaviorist and weather forecaster assigned to the incident. One important piece of information to be had from this communication is the relative humidity of the air. The primary reason is obvious: Lower humidity means increased fire spread. The ISO assigned to a structure fire needs to look at humidity for another reason: Humidity affects firefighters in a number of ways. In especially dry environments (hot or cold), firefighters become dehydrated quickly just by breathing. In high-temperature/high-humidity environments, the firefighter becomes dehydrated through profuse sweating in an attempt to cool the human machine (metabolic thermal stress). In cold temperature/high-humidity environments, the firefighter fights penetrating cold (pain) and the effects of rapid ice buildup on surfaces and equipment. High humidity also affects structural fires by keeping smoke from dissipating in the outside air. Prolonged low humidity can cause accelerated fire spread in lumber and other wood products, such as shake shingles and plywood, and flying brands can retain their heat longer and fly further in low humidity.

Temperature   Which is easier? Launching an aggressive brushfire attack when it is 100 degrees Fahrenheit or maintaining a critical defensive fire stream between two buildings when it's −10 degrees Fahrenheit? Answer: Neither. From the ISO's perspective, temperature needs to be evaluated relative to its effect on firefighter exposure. Acclimation is key! Firefighters working in International Falls, Minnesota, are well acclimated to the cold. A firefight in subfreezing temperatures may even seem commonplace. Conversely, firefighters in Yuma, Arizona, are probably accustomed to aggressive operations when it's 105 degrees Fahrenheit. Invert the two extremes and the stage is set for firefighter injury. The ISO needs to consider acclimation when evaluating the weather and its effect on firefighters. Practically speaking, you need to be concerned when your firefighters are operating outside their temperature "norm" for the region.

■ **Note**
From the ISO's perspective, temperature needs to be evaluated relative to its effect on firefighter exposure. Acclimation is key!

Potential for Change/Storms   In most cases, the weather prevailing for a given incident is what everyone will have to endure for better or for worse. At times, however, the duration of the incident or the development of a severe storm during it causes significant danger to responders. The ISO familiar with local weather tendencies can advise the IC of weather indicators that may require a shift in operations. Simple weather spotter guidelines and instructions, like those in the following list, can help the ISO warn others.

- *Watch the sky.* This simple activity is often overlooked. If foul weather is approaching or suspected, find a close vantage point to evaluate cloud patterns and wind activity. A quick phone call to mutual-aid agencies upwind may reveal useful information about the approaching storm.

- *Note 180-degree changes in wind direction in a short period of time (a few minutes).* Be especially wary of a sudden calm. A wavering smoke column is likewise noteworthy. In most cases, each event is a sign of an unstable air mass and may point to a significant weather change.

- *Be mindful of the potential for a flash flood.* Mentally compare previous rainfall and ground saturation with current rainfall to determine the potential for a flash flood. Evaluate your incident location and whether you are in a low, flood-prone area or along a drainage path.

- *Developing thunderstorms can produce rapid changes.* The changes include straight-line winds (micro bursts) of 100 miles per hour, hail, and lightning. Tornadic thunderstorms, characterized by a rotation between the rain-free base of a cloud and a forming wall cloud, are especially worthy of attention. A tornado forming in the vicinity of an incident or coming toward it should be immediately communicated to the IC and may be grounds for rapid building evacuation and a switch to defensive operations. Protective withdrawal from the area may even be considered.

- *At night, use lightning flashes to define cloud formations that may be tornadic.* Large hail (1/4 inch or larger) can also indicate that you are near the area where a tornado is most likely to form.

- *Calculate the distance between you and the lightning.* A simple rule of thumb to calculate lightning distance is to count the seconds between the lighting flash and the thunder. Divide this by five and you have the distance in miles. Thunder that claps less than 5 seconds from the flash means you are in the lightning area! Although lightning can strike anywhere at anytime, firefighters can protect themselves by becoming the smallest target possible. Stay away from poles, fences, and trees. Avoid standing in water. Suspend operations with aluminum ground ladders (practice solid risk management).

- *Deep snow not only makes travel difficult but can hide hazards at the incident.* Covered swimming pools, trip hazards, and electrical equipment can be hidden under snow. The weight of snow can stress buildings, as well as on utility lines, poles, signs, and other structures. Ice storms may cause the collapse of buildings, power poles and lines, as well as signage. Hose advancement becomes difficult in deep snow, causing the rapid fatigue of firefighters. Frequent crew rotation and rehab are key. Driven snow can impair visibility and make smoke reading difficult.

● Caution
Thunder that claps less than 5 seconds from the flash means you are in the lightning area!

Miscellaneous Hazardous Energy Forms

The list of hazardous energy forms can go on. Earthen materials (sand, gravel, stone, clay, and like materials), ice, flowing water (streams, canals, storm drainage), and even animal movement can present significant hazards to fire-fighters. The evacuation of penned, fire-threatened animals is amazingly dangerous. Have a plan. Some fire/rescue departments perform animal res-cues as a human life-saving activity (a deterrent to emotional and risky Good-Samaritan rescue attempts). It may seem humorous, but do not underestimate the power (hazardous energy) of a spooked dog, horse, cow, boar, bear, moose, llama, yak, wildebeest, or any animal. Medical care facilities have become increasingly high-tech with many devices, chemicals, and systems that can be included in the hazardous energy category. Magnetic resonance imaging (MRI) equipment uses highly energized magnets that can pose compelling hazards. As with all hazardous energy forms, be aware of their potential and expect the worst!

SUMMARY

Hazardous energy is defined as stored potential energy that causes harm if suddenly released. At an incident, the ISO needs to identify haz-ardous energy forms and categorize their poten-tial impact on firefighters. Forms of hazardous energy include electricity, utility gases and water and storm systems, mechanical energy, pres-surized vessels, vehicle components, weather, and other miscellaneous sources. Electrical equipment and systems are particularly haz-ardous, and ISOs need to know their particu-lars. Weather hazard considerations include wind, temperature, humidity, and the potential for storms and rapid change. Simple weather spotter guidelines and instruction can help the ISO determine the impact of weather as a haz-ard. Even animals can be a challenging form of hazardous energy during incidents.

KEY TERMS

Ground gradient   Electrical energy that has es-tablished a path to ground through the earth and that continues to energize the earth; an example is a downed power line.

Hazardous energy   Stored potential energy that causes harm if suddenly released.

POINTS TO PONDER

The Substation Fire

The alarm for an electrical substation fire came in around 1300 hours on one of those days when you would just rather stay indoors. The temperature was below zero, the wind was howling, and a stinging snow squall was just beginning. Upon arrival, crews found a large step-down transformer burning intensely. Exposures included the overhead power line grid, one other transformer, and a small control building. The local power company field representative had just arrived, and the decision to delay the fire attack was made. It took 10 minutes to verify that the substation grid was deenergized and power rerouted. The power company field rep then asked the fire department to start cooling the exposures and the transformer core. The IC, Battalion Chief Williams, and the designated ISO, Battalion Chief Ramirez, conferred on the request. The items they discussed included:

- There were adequate resources to launch the attack and an adequate water supply was nearby.
- The ground was frozen, and the drainage system built into the substation may not handle the cooling streams.
- The wind and temperature are likely to cause the stream spray to build up and form as ice on nearby high-tension wires, substructures, and towers. The structural integrity of these exposures was questioned.
- Failure to cool the intense fire may cause involvement of the second transformer or collapse of the overhead high-tension wires.
- Power has been disrupted for several hundred users.
- The weather was expected to get worse.

The discussion led to a planning session that involved additional representatives of the power company. During the session, Chief Williams began leaning toward launching the attack, whereas Ramirez leaned toward a delay to explore safer options. The power company representatives were equally split.

For Discussion:

1. Given what you know, are you leaning with Williams or Ramirez? Why?
2. What other factors would you like to consider in making an attack decision?
3. Should a "go or no-go" time frame be placed on the planning session?
4. What type of protective steps would you suggest for the firefighters once the operation commences?

Note:

This case study is taken from an actual incident. After an additional 20 minutes of planning and a 10-minute setup, the fire was attacked. Extinguishment took 40 minutes. Ice buildup caused one substructure to sag.

REVIEW QUESTIONS

1. Define hazardous energy.
2. List four ways to categorize the status of hazardous energy.
3. List common electrical equipment and its associated hazards.
4. What is a "ground gradient"?
5. Why are UPS battery rooms hazardous?
6. List the chemical properties of common utility gases.
7. List the hazards associated with utility water and storm sewer systems.
8. Give examples of mechanical hazardous energy.
9. What are some of the hazardous energy sources in vehicles?
10. What are the four considerations that need to be evaluated when considering weather as a hazardous energy?
11. List some warning signs that extreme weather is approaching.
12. List several sources of miscellaneous hazardous energy.

ADDITIONAL RESOURCES

Contact your local power and/or gas utility company for electrical and gas equipment hazard information. Many offer firefighter response training.

Linton, Steven J., and Damon A. Rust. *Ice Rescue.* Fort Collins, CO: Dive Rescue International, 1982.

National Weather Service. Web site containing many weather-related publications: Available at: www.nws.noaa.gov/om/publications.shtml.

Chapter 10

READING FIREFIGHTERS

Learning Objectives

Upon completion of this chapter, you should be able to:

- List the three factors that lead to overexertion, the three ergonomic factors that can produce injury, and the three strategies to mitigate ergonomic hazards.
- List the three factors that impact human cell performance, explain its chemistry, and define fuel replacement strategies to increase it.
- Discuss the two types of thermal stress.
- Explain the role of hydration in preventing injuries.
- Define the four Rs of firefighter rehabilitation.

READING FIREFIGHTERS—IS IT POSSIBLE?

In this section, we have focused on the need to front-load your knowledge so that you can rapidly apply knowledge at incidents through recognition and analysis (a skill). While the information in this section is far from complete, the intent is to provide a foundation for becoming an effective and efficient ISO. This last module on reading firefighters can be viewed as the wildcard because predicting firefighter behavior is not easy given the many factors that can impact firefighter judgment while performing tasks. Let's illustrate this.

Think of a response incident in your fire service career when an honest human mistake led to an injury or close call. At some point after the incident, you, in a moment of introspection, probably said, "What the heck was I thinking?" Most of us have many of such examples. Believe it or not, the answer to the question, "What was I thinking?" is simple: You weren't. Most working firefighter judgment mistakes can be explained—even those that defy explanation.

Human performance (including mental processing) is dependent on many factors. A breakdown in any of them can lead to decreased human performance—and mistakes. More often than not, overexertion causes mental (and physical) mistakes. We know that overexertion is the leading cause of injuries (and deaths) at incidents.[1] At the heart of reading firefighters is the evaluation of factors that lead to overexertion and therefore injuries. This chapter looks at the prevailing overexertion factors that help—and hurt—firefighters. In fact, the underlying assumption is that we, the fire service, have actually set up firefighters to be injured. This may sound condemning, but I think you will accept the statement if you agree with the setup.

The Overexertion Setup

■ **Note**
Firefighters are the only professional athletes who need to work at peak performance without warming up.

Consider the following: "Firefighters are the only professional athletes who need to work at peak performance without warming up."[2] Think about it, few occupations stress the human body to the degree that firefighters experience in hostile, working fire environments. Further, firefighters are called upon to perform such physical feats anywhere and at a moment's notice. Several celebrity athletes have even commented on the demands of the firefighter. The TV sports channel ESPN carried a special in which celebrity professional sports figures (some considered the most fit of all athletes) participated in a simulated firefighting task competition. Following the event, these athletes—to the person—admitted that the physical challenges of firefighting were off the charts. They couldn't imagine doing the same thing in smoke, heat, or mentally challenging situations. We all know that the incident scene is not a competition, although the race against time occasionally presents itself as such. This is but one part of the setup.

Most firefighters accept that an incident requiring peak performance can happen at anytime, although we structure our lives around the typical daily routine. Our chores, daily assignments, meals, workouts, and rest periods are mostly defined, and then we throw in an incident response when called upon. We accept this way of life, yet it is part of the setup. If you just ate, your body may not be able to operate at peak performance. If you are between meals, hungry, and the big one hits, you will struggle to perform at peak. If the alarm awakens you from a deep slumber . . . you get the point. These situations lead to rapid overexertion of the body and decreased performance of human cells, including brain cells! See, you weren't thinking!

Some believe that the fire service cannot change the reality that incidents occur while we sleep, when we are hungry, or when we are least capable of performing. The essence of fire service professionalism, however, is the ability to function safely at *all times.* Therefore, it is incumbent upon us to address the physical and mental demands on firefighters, regardless of *when* an incident occurs. Remember Chief Alan Brunacini's assertion: "For 200 years we've been providing a service at the expense of those providing the service." If we are truly dedicated to preventing injuries and deaths, we need to address the number one cause of injuries—overexertion. The solution for overexertion includes efforts to make sure firefighters can operate at peak performance.

Overexertion Resistance

The study of human performance and, more recently, of firefighter performance is well documented in texts, trade magazines, and medical journals. One conclusion is reached by all: Firefighting requires humans who are physically fit. Those who are not fit for duty have decreased overexertion resistance, and they present an injury risk to themselves as well as to others due to the team nature of firefighting. Conversely, the physically fit firefighter can resist overexertion longer, with minimal injury risk **(Figure 10-1)**. Nevertheless, both the fit and unfit firefighter can become fatigued, and regardless of their levels of fitness, both types of fatigued firefighters create an equal injury risk.

Enough cannot be said about the need for firefighters to proactively keep themselves strong, flexible, and aerobically fit. Further, they should adopt a program of efficient fueling for their metabolism. This preventive approach, in the long run, reduces injuries to firefighters. Firefighters who have strengthened their hearts to handle spike demands are obviously better off in this business. Those who know how their bodies respond to stress and how to immediately address metabolic deficiencies perform better. Again, this kind of preparation is all preventive, and all efforts to achieve this level of preparation helps in the overall firefighting effort. Still, incidents occur even when the most fit firefighter is not "warmed-up" or is not in a perfect cell-state to perform. That is the reality.

Figure 10-1 *The physically fit firefighter can resist overexertion longer with minimal injury risk.*

Overexertion Realities

The amazing diversity in physical performance capability (and incapability) of firefighters presents difficult challenges. At any given incident, some firefighters are operating at peak physical and mental performance, while others are ready to succumb to an overexertion injury. ISOs must deal with "here-and-now" overexertion threats to firefighters regardless of their preventive fitness efforts. ISOs can, however, make a difference in the way they address human overexertion factors. A solid understanding of overexertion factors can help ISOs quickly identify when the factors are adding to injury potential. With this foundation, they can help implement strategies to prevent overexertion.

The factors that influence overexertion are related to ergonomics, physiology, and rehab efforts.

■ **Note**
ISOs must deal with "here-and-now" overexertion threats to firefighters regardless of their preventive fitness efforts. ISOs can, however, make a difference in the way they address human overexertion factors.

ERGONOMICS

Some define ergonomics (which has only recently been applied to the workplace) as the science of adapting work or working conditions to a worker. Others define it as the study of problems associated with people adjusting to their work environment. In the fire service, both definitions are appropriate.

Figure 10-2   *Firefighters seldom find an "ergonomically friendly" workplace.*

The firefighter's workplace includes the fire station and apparatus (which can be engineered to be ergonomically friendly) and the incident scene (which creates challenges that are problematic). Firefighters don't have the luxury of having an "ergonomically friendly workplace" at an emergency scene! (See **Figure 10-2.**) However, we can minimize injury potential by understanding ergonomic stressors.

Ergonomic Stressors

Firefighters must work in an environment presenting ergonomic stressors that cause overexertion. Most of the activities at an incident require physical labor. Scientifically, this labor translates to human kinetics: muscular and skeletal motion. Being aware of the muscular and skeletal stresses and strains that firefighters undergo is the key to preventing overexertion. Often, the ISO needs to call a time-out and adjust the way firefighters have chosen to tackle the situation. To do this, the ISO should evaluate the environment, the

■ **Note**
The ISO should evaluate the environment, the relationship of the worker to the environment, and the task being attempted. Often, a slight change in any of these areas can reduce injury potential.

relationship of the worker to the environment, and the task being attempted. Often, a slight change in any of these areas can reduce injury potential. Each of these ergonomic stressors should be evaluated to determine if an injury potential exists.

The Physical Environment   This includes an examination of the surface conditions of the working area. Obviously, secure footing is essential to safe operations. Temperature variations, close proximity to equipment, and lighting integrity also need to be included here. Often, distractions are part of the work environment: noise, flashing lights, swinging mechanical equipment, and weather extremes.

The Relationship of the Worker   In a given environment, workers might be forced to tackle a task while bent down or stooping. Other times they must ascend or descend to accomplish a task. Pulling, twisting, and pushing to accomplish some objective are other relationships that may cause injury. The speed or pace at which the task is completed may also create an injury relationship.

The Task   The energy required to accomplish a task, as well as the amount of focus or attention required to complete it, are considered here. Sometimes, the task priority is such that additional stress is created. In other cases, the limited number of people available to accomplish the task is the primary factor leading to injury. The types of tools and equipment necessary to accomplish the task also need to be considered.

Ergonomic Abatement Strategies

Once the physical environment, the relationship of the worker, and the task have been evaluated, certain hazards become obvious. At that point, the ISO can utilize one of three strategies to abate or mitigate the hazard. These are awareness, accommodation, and acclimation.

■ **Note**
The ISO can utilize one of three strategies to abate or mitigate the hazard. These are awareness, accommodation, and acclimation.

Awareness   This strategy acknowledges that the workers are less apt to suffer an injury if they are aware of the problem, thereby heightening their cautiousness. Awareness is perhaps the most used abatement strategy and certainly the most simple. Examples include the warning of a slippery surface or a reminder to lift with leg leverage and a tight core.

Accommodation   Injury potential is reduced by altering the environment or the task. The use of personal protective equipment is accommodation. Placing a roof ladder on a pitched roof is a method of accommodating the worker in an environment that requires ascending and descending on a diagonal surface.

Adding more people to help perform a task is a form of accommodation. Artificial lighting is an accommodation to darkness.

Acclimation   This is the most difficult strategy to implement during an incident in that most acclimation is done proactively. Physical fitness programs that include strength and flexibility training for work hardening are examples of acclimation. One underused form of acclimation at an incident is to prehydrate crews and have them perform simple stretching exercises prior to an assignment. Other than first-due responders, this technique can prove to be very effective in preventing strains and sprains.

FIREFIGHTER PHYSIOLOGY

Physiological performance depends on the metabolic processing (the cell chemistry) of the firefighter. If the cell chemistry of a firefighter is not functioning optimally, the risk of overexertion increases. Factors that affect cell chemistry include thermal stress, hydration, and fuel replacement. Incident efforts to address these factors can be called rehabilitation, or "rehab." Let's explain each factor, then look at rehab strategies.

Thermal Stress

Thermal stress can come in the form of heat and cold. Heat stress can be further divided into internal (metabolic) heat and external (environmental) heat. The human body is amazing with its many systems to regulate a normal core temperature of 98.6 degrees Fahrenheit. Once the body temperature rises above this number, heat stress results. Factors that may cause core temperature to rise include:

- Activity
- Humidity
- Air temperature
- The effectiveness of cooling mechanisms
- Sun, shade, and wind

Heat Stress   Full structural protective clothing reduces the body's ability to evaporate heat by sweating (a cooling mechanism). Likewise, high humidity reduces the benefit of sweat evaporation. Working in direct sunlight adds additional heat stress, especially at higher altitudes (exposure to more ultraviolet rays). Firefighters suffering from heat stress go through of series of heat-related injuries that get progressively worse as heat builds. **Figure 10-3** outlines these heat injuries.

Heat Stress Injuries	
• Heat stroke	Medical emergency! Marked by hot, flushed, and *dry* skin.
• Heat exhaustion	Serious injury. Person is dehydrated, skin is cold and clammy. Person may be weak, dizzy, and nauseous.
• Heat cramping	Painful muscle spasms. Typically legs and arms are affected.
• Heat rash	May be an early warning sign.
• Transient heat fatigue	An early sign characterized by physical exhaustion. Remedied by rest and hydration.

Figure 10-3 *The ISO should be observant of the signs and symptoms of heat stress.*

The National Oceanic and Atmospheric Administration's heat stress index chart can help individuals predict heat stress injuries **(Table 10-1).** The chart is simple to use and should be referenced by the ISO when heat stress factors are present. If nothing else, it provides a reminder to accelerate crew rotation cycles and rehab efforts.

During structural fire operations, the ISO should *presume* that firefighters have elevated core temperatures. Typically, firefighters have used passive cooling to reduce core temperatures. **Passive cooling** includes the use of shade, air movement, and rest to bring down core temperatures. Paramedics and EMTs assigned to rehab functions have typically monitored firefighters' heart rates and their perceived comfort to determine when they have recovered sufficiently enough to don PPE and resume firefighting operations. New studies have shown that this approach is *not* adequate and that heart rate and perceived comfort are not good indicators of sufficient core temperature cooling.[3]

passive cooling
the use of shade, air movement, and rest to bring down human core temperatures

Core temperatures above 101 degrees Fahrenheit should trigger an active cooling strategy. **Active cooling** is the process of using external methods or devices (e.g., hand and forearm immersion, misting fans, gel cooling vests) to reduce elevated body core temperature. Several studies have shown that active cooling is best achieved using a forearm immersion technique. Firefighters doff their coats and submerge their hands and forearms in a basin of cold water. This technique is more effective than misting fans and has a reduced tendency to cause the chills (or sudden temperature change shock) that can occur when firefighters go immediately from hot environments to cold environments.

active cooling
using external methods or devices (such as hand and forearm immersion, misting fans, or ice vests) to reduce an elevated body core temperature

Cold Stress   Cold stress is similar to heat stress in that a series of injuries can occur if the body's core temperature cannot be maintained. In this case,

TABLE 10-1 *Heat stress index chart.*

HEAT INDEX											
ENVIRONMENTAL TEMPERATURE (F°)											
	70°	75°	80°	85°	90°	95°	100°	105°	110°	115°	120°
Relative Humidity	Apparent Temperature*										
0%	64°	69°	73°	78°	83°	87°	91°	95°	99°	103°	107°
10%	65°	70°	75°	80°	85°	90°	95°	100°	105°	111°	116°
20%	66°	72°	77°	82°	87°	93°	99°	105°	112°	120°	130°
30%	67°	73°	78°	84°	90°	96°	104°	113°	123°	135°	148°
40%	68°	74°	79°	86°	93°	101°	110°	123°	137°	151°	
50%	69°	75°	81°	88°	96°	107°	120°	135°	150°		
60%	70°	76°	82°	90°	100°	114°	132°	149°			
70%	70°	77°	85°	93°	106°	124°	144°				
80%	71°	78°	86°	97°	113°	136°					
90%	71°	79°	88°	102°	122°						
100%	72°	80°	91°	108°							

*Combined index of heat and humidity...what it "feels like" to the body.
Source: National Oceanic and Atmospheric Administration

APPARENT TEMPERATURE	HEAT STRESS RISK WITH PHYSICAL ACTIVITY AND/OR PROLONGED EXPOSURE
90° - 105°	Heat cramps or heat exhaustion possible
105° - 130°	Heat cramps or heat exhaustion likely, heatstroke possible
130° and up	Heatstroke highly likely
NOTE :	Add 10° F to the apparent temperature when working in structural PPE. Add 10° F to the apparent temperature when working in direct sunlight.

Cold Stress Injuries	
• Hypothermia	Can range from mild to severe. Mild cases are marked by shivering and loss of coordination. Lethargy and coma can onset quickly.
• Frostbite	A serious "local" injury meaning that a body part is frozen.
• Frostnip	A local injury. Most people do not realize they have frostnip. It is, however, a precursor to frostbite.

Figure 10-4
Hypothermia is the cooling of the body's core temperature, a condition that should be avoided at all cost.

however, the body temperature is being reduced or cooled. Moisture becomes an enemy. The heat and perspiration that firefighters experience while fighting an interior fire can cause shivering and a lack of concentration when they move outside to freezing weather. In these cases, the immediate temperature change sends a mixed signal to the hypothalamus. The body reacts, and one of the reactions can be decreased cognitive reasoning. Other factors that effect cold stress include air speed and temperature (wind chill), the level of activity, and the duration and degree of exposure. **Figure 10-4** outlines cold stress injuries.

Fighting Thermal Stress   The prevention of thermal stress injuries can be accomplished through accommodation, rotation, and hydration.

- Accommodation includes the use of extrawarm clothing during cold extremes or the use of forearm cold water submersion during extreme heat.
- Rotation is the planned action to rotate crews through rest, heavy tasks, and light tasks to minimize the stress caused by working in extreme environments.
- Hydration cannot be overemphasized in heat stress environments, but it is also effective, and often forgotten, in cold stress environments.

! Safety
The prevention of thermal stress injuries can be accomplished through accommodation, rotation, and hydration.

Hydration

Water is vital to the peak operation of virtually every body system from the transport of nutrients, to blood flow, to waste removal, to temperature regulation. When the body becomes dehydrated, these systems start to shut down to protect themselves. With shutdown comes fatigue, reduced mental ability, and, in extreme cases, medical emergencies such as renal (kidney) failure, shock, and death. Working firefighters must account for the wearing of PPE ensembles that do not allow for the evaporation of sweat, as well as for strenuous physical activity under mentally stressful situations. The hydration of firefighters should be paramount.

As a rule, firefighters should strive to drink a quart of water an hour during periods of work; this is best delivered in 8-ounce increments spread over the hour.[4] Substituting soda pop or other liquids for water can slow the absorption of water into the system. For this reason, just water should be given for the first hour. For activities lasting longer than an hour, some consideration should be given to adding essential electrolytes and nutrients to the water. Many sports drinks can achieve this. These sport drinks are best when diluted with 50 percent water to help speed their absorption into the system.

Fuel Replacement

While dehydration and thermal stress can lead to energy depletion, most firefighters associate energy depletion with the need for food. The nutrition, or "fueling," of firefighters can either rejuvenate them or put them to sleep. Too often, rehab feeding efforts accomplish the latter effect. The ISO should view firefighter fueling from the perspective that a properly nourished firefighter works smarter and safer. The *improperly* fed firefighter not only wants to "crash," but likely makes sluggish mental calculations, leading to injury. (Remember that question, "What was I thinking?")

So what is the proper way to nourish firefighters? A study of essential nourishment theory can answer this question. (Remember, however, that individual metabolisms are diverse.) Metabolic rates are influenced by lifestyles, fitness, illnesses, over-the-counter and prescription drugs, and circadian rhythms. **Circadian rhythms** reflect a person's physiological response to the 24-hour clock, which includes sleep, energy peaks, and necessary body functions. The following explanation of cell theory and strategy is meant to maximize cell performance for hardworking responders at an incident.

circadian rhythms
a person's physiological response to the 24-hour clock, which includes sleep, energy peaks, and necessary body functions

Cell Theory   In basic terms, the physical and mental performance of a firefighter depends on each and every cell working optimally. For a cell to work as well as it can, it must use a balance of four essential elements: oxygen, water, glucose (from food), and insulin. When this balance is present, other essential hormones and enzymes are created that make for a well running human machine. The often misprescribed element is food: the foundation for building balance. True, *all* food is fuel for the firefighter. That is, all carbohydrates (carbs) from food are converted to glucose when the muscles and brain are being worked hard. Insulin is released into the blood stream to help regulate sugar use. The entry rate of a carbohydrate (glucose) into the bloodstream is known as the glycemic index. A high-glycemic-index food source quickly spikes blood sugar (and the resulting insulin surge). The sugar is rapidly burned at the cell level, leaving too much insulin in the system, which is telling the body to store carbs, not use them. This is dangerous! It makes sense, then, to fuel firefighters with a low glycemic index food so that the blood sugar levels and insulin are stable, gradual, and consistent (**Table 10-2**).

TABLE 10-2　*Glycemic index comparisons for common carbohydrates.*

High Glycemic Index Carbohydrates (to Avoid)

White sandwich bread	Pizza crust
Hamburger/hot dog buns	Durum spaghetti
Rice cakes	French fries
White flour tortillas	Chips/pretzels
Orange juice	Most cereals
Watermelon	Cookies
Grapes	Cakes
Raisins	Donuts
Bananas	Bagels
Potatoes	Pastries
Carrots	Macaroni
Pumpkin	Most candy
Sweat Corn	

Low to Moderate Glycemic Carbohydrates (Better Choices)

Most vegetables	Noninstant oatmeal
Apple juice	Oat bran breads
Apples (whole)	Mixed whole grain breads
Oranges (whole)	Whole grain tortillas
Pears	Low-fat yogurt
Peaches	Soups without pasta
Most beans	Most nuts
Milk	

The glycemic rate is but one part of the equation. The cell-fueling strategy must also include other essential elements to keep energy levels high.

Safety

The key to providing quick energy that optimizes cell performance is to feed firefighters a balance of low glycemic carbohydrates, protein, and fat.

Cell-Fueling Strategy　The key to providing quick energy that optimizes cell performance is to feed firefighters a balance of low glycemic carbohydrates, protein, and fat. Ideally, this balance should be a 40/30/30 mix, that is, 40 percent carbohydrate, 30 percent protein, and 30 percent fat.[5] This balance provides essential elements from three food groups. A balanced approach achieves the following benefits:

- The low glycemic carbs stabilize insulin release into the bloodstream, helping to reduce the roller coaster of blood sugar levels that often lead to sporadic activity, chemical imbalance, and fatigue.

- The protein helps cells rejuvenate and facilitates the building of new cells (amino acids are the building blocks of cells).
- Dietary fat helps essential hormones form and stabilizes the carbohydrate entry rate. Additionally, dietary fat can signal the body that "enough" food has been eaten and helps limit overeating by working crews. Overeating can tax the digestive system, causing blood to shift from the muscles and the brain; this is why we sometimes feel like sleeping after a meal.

Choosing the best protein, carbohydrate, and fat also promotes steady, sustained performance. Protein is best derived from low-fat meats such as turkey, chicken, and fish. Eggs and cheese also offer protein. Fats should be of the monounsaturated type like olive oil, nuts, and low-sugar peanut butter. Carbohydrates are the toughest to balance correctly because so many of the foods we typically find on the fire scene are rich with unfavorable carbohydrates. Donuts, fries, candy, bread, potatoes, bananas, fruit juice, and chips/crackers all have moderate to high glycemic indexes. Good carbohydrates include green vegetables, apples, whole-grain breads, tomatoes, and oatmeal.

The department that preplans nourishment for rehab efforts can achieve the desired balance. Menu planning with a nutrition or fitness expert is one way to preplan rehab. When a department has not preplanned rehab nourishment, the ISO can make recommendations based on the balance approach. A loose sampling of nourishment for rehab is included in the Sampling of Balanced Nutrition for Firefighters Working at an Incident text box.

Sampling of Balanced Nutrition for Firefighters Working at an Incident

Morning: Breakfast burritos, apples, and water. Breakfast burritos can be made with scrambled eggs (egg white only is best), chopped Canadian bacon, and low-fat cheese, all wrapped up in a whole-grain flour tortilla. Fresh salsa can also be added.

Afternoon/evening: Turkey and cheese sandwich (on thinly sliced whole-grain bread), water, apples, and small portions of peanuts or cashews. It is better to have high-piled turkey than to eat two or three sandwiches; the increase in bread intake eventually throws off the balance. Mustard will not hurt the balance, whereas catsup and mayonnaise can.

Anytime: Commercially available energy bars and water. It is important to make sure that the energy bars have the 40/30/30 balance or state they are "low glycemic index."

Eating Cycle at Incidents   Deciding *when* to feed firefighters is subject to debate. One solution is simply to fuel the firefighters when they are hungry. This aim sounds simple, although its application may be subjective and reliant on perception. An understanding of theory provides good planning insight into when to feed firefighters.

When a firefighter is working hard, cells pull glucose from the bloodstream, which the bloodstream gets from the liver. The liver has a limited amount of readily available glucose. When the glucose is used up, the liver looks to pull glucose from the digestive track. If it has been more than two or three hours since food was ingested, no readily available carbohydrates are available. In these cases, the body attempts to gain glucose by breaking down muscle or by starting the long, difficult process of breaking down body fat stores. In either case, undesirable by-products are created that interfere with optimal cell performance (time, more water, heat, and lactic acid). From this process we can derive a good rule of thumb for feeding firefighters:

- Feed *now* if it's been more than two hours since the last food intake.
- Feed *every two to three hours* when physical and mental demands remain.

Granted, the more physical the task is, the more important feeding becomes. Often, fire department members performing less physical tasks (incident commanders, ISOs, staging managers, and others with such assignments) get left out of the fueling cycle. Remember, optimal thinking requires optimal cell performance also. Those assigned to less physical tasks require balanced fueling—perhaps at a lower *volume* than those physically working hard.

REHABILITATION EFFORTS

The rehab effort that includes balanced nutrition, substantive hydration, and relief from thermal stress keeps firefighters performing well, both mentally and physically. This effort, in itself, reduces mental lapses and injury potential caused by overexertion. The NFPA 1521 standard addresses rehab as a function of the ISO: The incident safety officer shall ensure that the incident commander establishes an incident scene rehabilitation tactical management component during emergency operations. It is not the task of an ISO to manage and perform rehab, only to make sure the IC has addressed this important function. The ISO should take this a step further and evaluate the rehab efforts to make sure that they are being effective. To evaluate rehab efforts, the ISO needs to know the components of effective rehab, which should include the four Rs: rest, rehydration, Rx, and refueling (see The Four Rs of Rehab textbox).

The Four Rs of Rehab

Rest: This is a time-out to help firefighters stabilize their vital signs. The most important vital sign to stabilize is core temperature (which should be less than 100 degrees Fahrenheit and greater than 97 degrees Fahrenheit). Relief from thermal stressors is essential; active cooling is suggested for elevated core temperatures. Pulse rate, blood pressure, and breathing rate should return to normal. Rehab attendants should not rely on "perceived comfort."

Rehydration: This includes the replacement of fluids and stroke volume lost to perspiration and muscle activity. Water is the primary fluid to replace. Electrolytes need to be replaced if the firefighter has been sweating for more than an hour.

Rx: This stands for medical monitoring and treatment. Paramedics (preferably) or EMT-basics should make a judgment on whether a firefighter can return to incident duties based on their best judgment and the vital signs acquired from the individual firefighter.

Refueling: Make sure that provisions are made for balanced food nourishment to improve sustainable energy and mental acuity.

It is not the duty of the ISO to help rehab attendants accomplish their task. The ISO should offer suggestions when an essential part of rehab is not being addressed.

SUMMARY

The ISO needs a foundation to read firefighters. Most firefighter mistakes and injuries at an incident can be attributed to the onset of overexertion, which is triggered by ergonomic and physiological stressors. The ISO who understands the underlying theory of stressors can better prevent overexertion. Ergonomic stressors fall into the categories of the physical environment, the relationship of the worker to the environment, and the task being performed. Abatement strategies for ergonomic hazards include awareness, accommodation, and acclimation. Physiological stressors are influenced by thermal stress, hydration, and fuel replacement. Thermal stress comes in hot and cold forms, and either form can lead to life-threatening injuries. Abating thermal stress is achieved by crew rotation, accommodation (such as active cooling), and hydration. Optimal human performance relies on optimal cell performance. Understanding the theory and practicing the strategies for optimal cell fueling can reduce injuries and mental lapses. Balancing food intake and eating in regular cycles are essential. One of the ISO's functions is to make sure rehab efforts are being addressed. The ISO should also evaluate rehab efforts to make sure the four Rs are implemented and achieving results.

KEY TERMS

Active cooling   Using external methods or devices (such as hand and forearm immersion, misting fans, or ice vests) to reduce an elevated body core temperature.

Circadian rhythms   A person's physiological response to the 24-hour clock, which includes sleep, energy peaks, and necessary body functions.

Passive cooling   The use of shade, air movement, and rest to bring down human core temperatures.

POINTS TO PONDER

The Dump Fire

Lieutenant James and Driver/Firefighter Chan made up a two-person engine crew assigned to the outlying station of a combination fire department. They had a productive morning that included their vehicle readiness checks, usual morning workout, a vehicle extrication incident, and the weekly task of mowing the grass in front of the station. The summer day was typical enough, although James and Chang decided to take a break from the heat and address their lunchtime hunger pangs. Just as they sat down, the station alarm alerted them to a single-engine response to a rubbish fire. Upon arrival at the scene, they found a small deep-seated fire in an illegal dump down in a ravine. They reported that they could handle the incident.

Training/Safety Officer O'Connor was busy in her office, casually monitoring the radio while James and Chan worked the dump fire. A series of radio transmissions over the next 20 minutes caught her attention. The first asked for a volunteer manpower response. The next asked for a county backhoe to assist with digging out the trash. Finally, O'Connor heard a request for an EMS response for an injured firefighter—which came from Chan—not James. O'Connor grabbed her radio and responded to the dump in accordance with her responsibilities as safety officer.

Upon arrival, O'Connor was informed that Lieutenant James had experienced a blackout and was being treated with IV therapy in the back of the ambulance and would likely be transported to the hospital.

For Discussion:

1. What ergonomic factors contributed to Lieutenant James's injury?

2. What thermal stress factors contributed to this incident?

3. From a department operational viewpoint, what traps were present?

4. What interventions could James, Chan, or even O'Connor have employed to prevent the outcome?

Note:

This case study is taken from an actual injury investigation report. Names were changed to protect the guilty!

REVIEW QUESTIONS

1. List the three factors that lead to overexertion.

2. What are the three ergonomic factors that can produce injury?

3. What are the three As to help mitigate ergonomic hazards.

4. List the three factors that impact human cell performance.

5. Discuss the two types of thermal stress.

6. List three examples of passive cooling and two methods of active cooling.

7. At minimum, how much water should working firefighters drink at an incident?

8. What four elements need to be balanced to help human cell performance?

9. When feeding firefighters, food should be geared toward what balance?

10. How often should firefighters eat when incident activities require significant effort over a long period of time?

11. Define the four Rs of firefighter rehabilitation.

ADDITIONAL RESOURCES

NFPA 1584, *Rehabilitation of Members Operating at Incident Scene Operations and Training Exercises,* 2003 ed. Quincy, MA: National Fire Protection Association, 2003.

Ross, David, Peter McBride, and Gerald Tracy. "Rehabilitation: Standards, Tools, and Traps." *Fire Engineering* (May 2004).

Sears, Barry, PhD. *Enter the Zone.* New York: HarperCollins, 1995.

NOTES

1. Firefighter injury and death reports from the NFPA, USFA, and IAFF unanimously conclude that this statement is factual.

2. Variations of this statement have been presented by numerous fire service speakers over the past several years. The original source is not known.

3. Thomas McLellan, PhD. *Safe Work Limits While Wearing Firefighting Protective Clothing,* Toronto Fire Department Grant Study Report (Toronto, ON: Defense Research & Development Council, 2002).

4. United States Fire Administration, *Emergency Incident Rehabilitation*, FA-114 (Washington, DC: USFA Publications, July 1992).

5. Barry Sears, PhD, *Enter the Zone* (New York: HarperCollins, 1995).

THE ISO ON-SCENE

FEELING THE PINCH

Ever have one of those days when you felt like you were caught between two warring factions and you struggled to find middle ground while upholding safe principles? I experienced such a day a while ago—and the lessons I learned from that event are included in many of the chapters of this section.

At the time, I was assigned as the training officer and HSO for my small combination department. While in my office, I heard an engine and duty fire officer being dispatched to a single-patient medical call at a public elementary school for a fall victim. The duty officer arrived and established command, followed by the engine. A bit later, the engine was released, but the IC reported that he was staying on-scene, awaiting a building or contractor representative (school was out of session for the summer). About ten minutes later, dispatch called my office and said that something weird was going on: They had received numerous calls from the media concerning the school collapse and the public information officer (PIO) could not be reached (I had occasionally filled that role when the PIO was not available). I immediately called the IC, who reported that a portion of the school's roof had collapsed, a construction worker had been briefly pinned under it, but the engine crew made a quick rescue. I informed the IC that we were getting several media requests and that I would gladly come out and help until the PIO was located.

On arrival, I could see that a construction crew was in the middle of installing a new roof on the school and that a stack of materials placed on the roof had overloaded some trusses and caused a collapse. I also noted six or seven other material stacks on the remaining roof. An examination underneath revealed that several partition walls were bowed out, trying to support the ends of some of the lightweight steel trusses that had collapsed. I shared my fear that additional collapse potential was high, not only for the portion that had partially collapsed but also under the other material stacks. The IC acknowledged the probability but said it was the contractor's problem. The construction workers were awaiting the arrival of their supervisor but were intent on returning to the roof and beginning damage cleanup. My life safety fears kicked in and I decided to talk this out a bit with the IC—who finally concurred that we should place the building off-limits and call a building inspector. The IC requested an engine to come back to the school for standby and to help secure utilities. The power and gas companies were also notified.

The pinch got worse. The contractor supervisor arrived and wanted to get his crews back to work. He was obviously angry and kept trying to downplay the seriousness of the problem. The IC was starting to give in. A building inspector arrived but would not take a side saying that the school's structural engineer should be involved. A school district representative arrived and agreed with the contractor. It was then that I thought the IC would turn over the incident. I encouraged the IC to contact the chief's office and get his opinion on our liability in turning over the building (the chief said to keep control). Eventually, we had structural engineers on-scene who disagreed on the problem (representing the school district and the roof contractor), building inspectors who disagreed on condemnation, and school district officials who were looking to the fire department for guidance on the safest way to eliminate the threat.

I was assigned by the IC to be involved in the planning process and decided to take the approach that we do nothing until we agree on a plan. Representing the fire department, I stood by the need to have the damaged roof shored before any debris removal and to have the other materials removed by crane—but the crane technician needed to be harnessed and tethered from above in case the roof collapsed further while attaching the crane sling. It took every diplomatic bone in my body to keep focused on a safe approach while engineers, contractors, and stakeholders argued about who was responsible and what should happen next. Finally, the school superintendent stepped to the plate and empowered the fire department and the school's retained structural engineer to come up with a shoring plan and supervise the material off-loading. I was reassigned as the ISO to monitor the fire department shoring operation, followed by the rope-harness belay from our aerial device, while the other material stacks were being removed by the crane.

The next day, I was contacted by the fire chief and asked to conduct a thorough review of the incident and document all of our discussions, actions, and issues for the sake of a formal postincident critique—and for the school district's attorney who was already preparing for cost-recovery issues. My investigative and writing skills were put to the test and were critiqued by our city attorney and the school district's attorney. The contractor's law firm challenged many things in my report. Later, I was prepped for court testimony but never got subpoenaed.

I hope this section of the book gives you some tools to help when you feel the pinch.

Dave Dodson

Chapter

11

TRIGGERS, TRAPS, AND WORKING WITHIN INCIDENT COMMAND SYSTEMS

Learning Objectives

Upon completion of this chapter, you should be able to:

- List four methods that will help the ISO trigger safe behaviors.
- List the three ISO "traps" and discuss how each can render the ISO ineffective.
- Describe the organizational position of the ISO within the ICS.
- List the two primary communication tools the ISO uses and list guidelines for each.
- Define the national "typing" scheme and how the ISO function can expand for small and large incident types.

INTRODUCTION

While the front-loading (described in Section Two) can help the incident safety officer become technically competent, by itself it does not create an ISO who will make a difference. The element that tips the scale toward making a difference is the ability to communicate clearly and appeal to the safety sense that can be sidestepped during working incidents.

Firefighters are by and large a proud, strong, and reasonable group who thrive on competition, adrenaline, and challenge. The ISO can trigger favorable or unfavorable responses when confronted by a group of firefighters driven by challenge. Often, the new or inexperienced ISO falls into some common traps that may seem trivial, yet they thwart the ultimate goal of making the incident safer. Further, the ISO who does not address the traps becomes only minimally effective in the long-term. Likewise, the ISO must present concerns in a way that is appropriate and appealing for the incident commander. The failure to work in harmony with the IC is also a failure in firefighter safety. This chapter looks at ways to trigger favorable results and avoid harmful traps. Additionally, we look at ways to work within incident command systems and improve your relationship with the incident commander.

■ **Note**
The ISO can trigger favorable or unfavorable responses when confronted by a group of firefighters driven by challenge.

■ **Note**
The failure to work in harmony with the IC is also a failure in firefighter safety.

TRIGGERS

The incident scene is much like a championship football game in that everyone present is trying to score a goal. To get the win, a coach must develop a strategy for victory and assign tactics (plays) to meet the strategy, while harnessing the varying skills, talents, fortitude, and emotional levels of the players. While everyone wants to win, occasionally a player operates outside the coach's plan—usually resulting in a loss.

When fighting a fire, the incident commander utilizes crews who possess a variety of skills, knowledge, talents, and emotions. Occasionally, a player (firefighter) operates outside the plan. The firefighter wants to "win," but the emotion, skill, talent, or method in play is not appropriate for the incident commander's plan. Unfortunately, the loss is not just of a game; it is likely to be the loss of life or the loss of a working firefighter to injury. This is where the ISO can step in and make a difference. By using certain "triggers," the ISO can remind individuals that a professional firefighter operates according to a plan, in a safe manner, and within solid risk/benefit ranges. *Work safe triggers* can be defined as the basic approach of an ISO to help firefighters work more safely. Some triggers are subtle (passive), whereas others are pretty direct (active). Let's start with the subtle and work toward the more direct.

Figure 11-1 *Be visual. The use of a SAFETY vest can help trigger safe behavior by everyone. (Photo courtesy of Richard W. Davis.)*

Visibility

The incident safety officer should wear a high-visibility vest that clearly states "SAFETY" (**Figure 11-1**). While this may seem like a simple, passive trigger, it is surprisingly effective. In many ways, this is like the "power of suggestion" employed by many advertisers. Upon seeing the safety vest, fire-fighting crews often stop an unsafe action or withdraw from an unsafe position. This self-correction is desirable on-scene; the ISO can then concentrate on other items of concern. This trigger may seem like an elementary or even childish suggestion. Experience, however, shows that this trigger is quite effective, and therefore a useful tool for the ISO.

To maximize this trigger, the SAFETY vest should be instantly recognizable from a distance. Many vests look alike from a distance, and it can be difficult to differentiate the ISO vest from the staging manager or even incident commander. As stated in Chapter 4, using a different color and highly reflective trim seems to be the best solution for safety officer vests. While no standard exists, most people associate the color green with safety. The National Safety Council uses a green cross as their symbol and the Fire Department Safety Officer's Association uses a green Maltese cross as its symbol.

Example

Philosopher, theologian, and Noble Peace Prize winner Albert Schweitzer once said, "Example is not the main thing in influencing others. It is the only thing."[1] Effective ISOs always try to set a good example while performing their duties. Often, it is the habits and self-discipline that an incident safety officer displays that influence others—a passive trigger to safe behavior. To illustrate, ISOs should always, without fail, participate in the crew accountability system. Likewise, they should use appropriate PPE, follow department policies, and obey zone markers.

Because ISOs often work alone—contrary to most fire service tenets—they are usually outside the hazard area and looking in, and this practice is acceptable. However, to set a good example, ISOs working alone should follow basic guidelines:

- Always be in sight of another responder.
- Always be within shouting distance of another responder.
- Let somebody know where you are going if you are taking a tour of the incident scene.
- Don't walk into, or breathe, smoke.

These guidelines may seem a bit absolute and impractical, but their intent is to set the example. It is counterproductive for the ISO to be viewed as working unsafely. In fact, many firefighters find it fun to catch the ISO acting unsafe—and the teasing can be relentless. Do not give them the chance. ISOs should evaluate their own environment and exposure, making the adjustments necessary to act safely. Here are some examples:

- In the unusual case where ISOs need to go into an IDLH environment or into a hot zone, they should request a partner, use appropriate PPE and SCBA, and be tracked just like any other assigned crew.
- When performing reconnaissance around a building, ISOs are likely to walk into an area where no responder is visible (if you cannot see anyone, then nobody can see you). In these cases, ask for a partner to go with you. If no one is available, walk to the bravo-charlie corner, then walk back around the A-side to see the charlie-delta—and have someone on the A-side keep you in their vision.
- ISOs need to self-monitor their rehab needs: Stay hydrated and eat something if you have been on-scene longer than two hours. Take steps to minimize the effects of thermal stress on your thinking ability.

■ **Note**
A soft intervention can be defined as an effort to make crews, command staff, and general staff aware that a hazard or injury potential exists.

soft intervention
an intervention to make crews, command staff, and general staff aware that a hazard or injury potential exists

Soft Intervention

Being readily identifiable and setting a good example are subtle triggers that can be classified as passive. To *actively* address safety issues, the ISO needs to intervene in some way. Safety issues requiring intervention can be classified as imminent threats or potential concerns. If the threat is *imminent*, immediate intervention is required: A firm intervention (discussed later). All other concerns can usually be addressed by soft intervention. A **soft intervention** can be defined as an effort to make crews, command staff, and

Figure 11-2 *Often a simple reminder (a soft intervention) is all that is needed to prevent an injury.*

general staff aware that a hazard or injury potential exists (**Figure 11-2**). The intent of this awareness is to achieve positive safety changes in behaviors, operations, and actions within the framework of the incident management system and the incident action plan. The use of humor, subtle reminders, information sharing, and "peer-talk" are examples of soft interventions. These are employed as a way to trigger safe behavior when the witnessed action or environment is not necessarily life-threatening. In most cases, the person receiving the message makes adjustments to the hazard within the framework of IMS. This is especially true if the ISO acknowledges the wisdom or choice made previously, then interjects a third interpretation or additional information that underscores the safety concern.

Interjecting humor is effective when making soft interventions—but be careful with it. The humor must not trivialize the safety concern. To make a safety point, humor is best used when the environment allows face-to-face communication and centers on the circumstance, not on the actions of firefighters. Self-depreciating humor is rarely offensive to others and can be effective.

Soft intervention should not be used to stop, alter, or suspend actions or operations. The reasons are obvious: First, most recognize that the ISO can stop, alter, or suspend activities and operations only if an imminent threat to firefighters is present. Second, the next time a threat *is* imminent, its seriousness must not be misinterpreted.

Firm Intervention

firm intervention
an intervention to immediately stop, alter, or suspend an action or operation due to an imminent threat; more or less an official order to stop, alter, or suspend an action

imminent threat
an activity, condition, or inaction that will most certainly lead to injury or death

Firm intervention can be defined as actions to immediately stop, alter, or suspend an action or operation due to an imminent threat. The firm intervention is equivalent to an official order to stop, alter, or suspend an action. In these cases, you are exercising the authority of the incident commander. An **imminent threat** is best defined as an activity, condition, or inaction that will most certainly lead to an injury or death (**Figure 11-3**). For example, an ISO who witnesses a crew operating in an imminent collapse zone could relay via radio, "Attack Team 3, from Safety. Evacuate that position immediately. That is an imminent collapse zone. Acknowledge." Anytime a firm intervention is used, the ISO should immediately relay the concern to the incident commander. The authority to use and report firm interventions is reflected in the NFPA 1521 Standard (see Authorities of the ISO from NFPA 1521).

> ■ **Note**
> **Anytime a firm intervention is used, the ISO should immediately relay the concern to the incident commander.**

Authorities of the ISO from NFPA 1521

The authority and requirement to use and report firm interventions can be found in the NFPA 1521 standard:

4.6.2   At an emergency incident where activities are judged by the incident safety officer to pose an imminent threat to firefighter safety, the incident safety officer shall have the authority to stop, alter, or suspend those activities.

4.6.3   The incident safety officer shall immediately inform the incident commander of any actions taken to correct imminent hazards at the emergency scene.

Figure 11-3   *Firm intervention is used to stop, alter, or suspend activities that pose an imminent threat to firefighters. (Photo courtesy of Richard W. Davis.)*

Realistically, hazards and corrective needs at an incident may fall between the need for a soft and a firm intervention. In these cases, the ISO should try a soft intervention first. If the soft intervention is ineffective, the ISO may choose to use a stern advisory. In essence, the ISO is stating in clear, direct language that the hazard or behavior is most alarming. The use of a stern advisory should be reserved for the few unfortunate occasions when an individual or crew is acting without discipline or with a perceived disregard for the safety of themselves or others. Matching the intervention to the degree of concern is essential to achieving buy-in with the person or crew in question. If the intervention is viewed as irrelevant or demeaning to the crew, change may not occur. When the stern advisory does not achieve change, consider taking the issue (and solution) up the chain of command.

TRAPS

Incident safety officers can find themselves getting trapped into operational modes and activities that render them (and the ISO program) ineffective. Specifically, responders dismiss the ISO's general approach. Often, it is the well-intentioned, inexperienced ISO who wanders into the following types of traps.

The Bunker Cop

The *Bunker Cop* syndrome is evident in the ISO who spends too much time looking for missing, damaged, or inappropriate use of personal protective equipment. For example, the ISO may be always asking working firefighters, "Where are your gloves?" or "Why isn't your helmet on?" Like a traffic cop, the Bunker Cop is focused on one specific component of safety: protective equipment.

Unfortunately, this focus causes the ISO to miss the big picture of incident safety (**Figure 11-4**). Further, it eventually alienates the ISO and undermines an ISO program in the eyes of firefighters; most firefighters know what the PPE expectations are and they resent the "babysitting." Occasionally, firefighters need to be reminded of safety expectations, but the reminder is best delivered by the company officer, team leader, or group supervisor. If these leaders take care of PPE issues, the ISO does not have to be on patrol for them.

On the other hand, ISOs could be considered negligent if they fail to recognize situations in which firefighters are not utilizing appropriate PPE. In these cases, they should report the infraction to the person's team supervisor as a soft intervention. If PPE issues keep resurfacing at incidents, the department values and SOP enforcement are delinquent. If so, the ISO needs to work within the framework of departmental change and, unfortunately, is forced to be a bunker cop in the short term.

Figure 11-4 *The ISO who takes the Bunker Cop approach misses the big picture.*

Another type of Bunker Cop focuses on firefighter skill proficiency at incidents. It is easy to watch a crew throwing ground ladders and determine whether they are doing it "by the book," but the ISO who brings up skill deficiencies may not be received well by working crews. An ISO once made this mistake: The senior officer who was operating as a company officer found the ISO intervention very annoying, promptly put the ladder down, and said, "Here, if you're so damn smart, you do it!" To avoid this predicament, ask yourself what is the likelihood of an injury if they continue and how serious will the injury be? (Remember risk management in Chapter 2.)

New or inexperienced ISOs often fall into this trap because of their familiarity and comfort with basic PPE use and skill evolutions. When the new ISO is (or is acting as) a company officer, PPE and crew safety are comfort zones, and nobody on the crew questions the company officer's authority to address the issues. While it may feel uncomfortable to do so, the new ISO needs to get past PPE and skill issues and look at big-picture items, such as fire behavior, building construction, and risk management.

The CYA Mode

The well-known abbreviation *CYA* best describes incident safety officers who spend an inordinate amount of time ensuring that they are not held personally accountable for incident scene actions. Although the label is earthy, the behavior must be acknowledged. The proliferation of NFPA standards, OSHA Code of Federal Regulations (CFRs), and local requirements have placed ISOs in a position of liability by mandating their compliance with the due diligence requirements outlined in the codes and standards. It is easy to imagine an incident safety officer being found liable for a significant firefighter injury or duty death if he or she did not do meet requirements. In fact, a case can easily be made that no death should occur if an incident safety officer is on-scene. This environment can lead to a CYA approach by the ISO.

In a CYA approach, the ISO is constantly citing CFRs, standards, and other numbered requirements as the reasons for bringing up safety concerns: "You can't do it that way because OSHA 1910.134 says so." A worst case is one in which the ISO tries to "wash his hands" of a safety infraction. Usually, this is a result of an ISO's inability to get an incident commander or other officer to change an operation or task: "If someone gets hurt, it's not my fault."

In all cases, ISOs who use CYA tactics are destined to fail. Firefighting crews see right through their interventions and simply dismiss their recommendations or orders as pure self-preservation. To avoid this outcome, ISOs must display a genuine concern for everyone's safety. They must take personal responsibility for each and every firefighter's safety. While codes and regulations may tell them safe ways of doing things, they meet requirements not because of the requirement, but rather because they believe in doing things the safe way—to reduce the threat of injury. To avoid the CYA label, practice "good intent" and "personal concern."

The Worker

The ISO who pitches in and helps crews with their tasks falls into the *worker* trap (**Figure 11-5**). To be effective, ISOs must stay mobile and observant; they must not allow themselves to get involved assisting with a fixed assignment. Too often, this is easier said than done. In many cases, ISOs may feel it is necessary to help move hose lines or throw ladders because there simply are not enough people to do all the assigned tasks. If this is the case, the incident commander should consider retaining the safety officer responsibility and allow the former ISO to lead and work with an assigned crew. Perhaps more effectively, the tasks and risks should be reevaluated and prioritized to match the number of the persons available. If there are too few people to do assigned tasks, then the ISO role becomes even more critical, and the need to see the big picture and not get caught up in activities are all the more important.

Before his retirement, Division Chief Gene Chantler of the Poudre (Colorado) Fire Authority said it best: "If an officer assigned as SO [safety officer] becomes directly involved with suppression . . . there are two people who made a mistake. The first is the individual assigned to the position; it is clear they are unsuited to fill that role. The second person to make a serious mistake is the IC for appointing someone to the position who doesn't clearly understand what the job entails."[2]

These three traps can, and have, wrecked many incident safety officer programs. The ISO must look past them and be constantly vigilant to avoid them.

Figure 11-5 *The ISO must be careful not to get suckered into performing tasks. (Photo courtesy of Richard W. Davis.)*

WORKING WITHIN COMMAND SYSTEMS

The effective and efficient ISO must work within the framework of an established incident management system (IMS). Organizationally, the ISO typically reports directly to the incident commander, even though there are still variations of IMS within the fire service. Nationally, efforts are being made to unify the various incident management systems (NIMS [National Incident Management System] compliance) so that all jurisdictions will use a common language and expansion processes. With this in mind, we need to address a couple of issues: (1) incident commander relations, (2) the national incident typing scheme, and (3) the expansion of the ISO function in incidents where a single ISO cannot perform all the required duties.

Incident Commander Relations

Virtually all incident management systems hold the incident commander responsible for the safety of responders. This responsibility may not be "dished out" to an ISO; the IC is ultimately responsible. In the section dealing with ISO authorities, the NFPA 1521 standard addresses responsibilities by stating: "At an emergency incident, the incident commander shall be responsible for the overall management of the incident and the safety of all members involved at the scene." When utilizing an ISO, the IC is putting

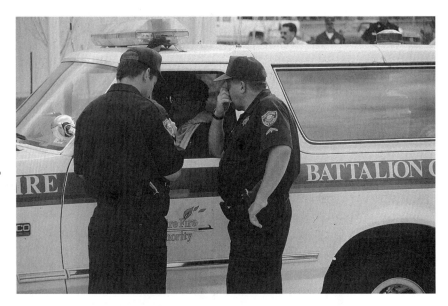

Figure 11-6 *The ISO must respect the fact that the ultimate authority for firefighter safety rests with the incident commander.*

trust in the ISO to handle the safety function as an extension of the IC's authority (**Figure 11-6**).

Making a difference as an incident safety officer is codependent on the ISO's support of the incident commander and an IC's faith in the ISO. When the two elements come together, the injury and death potential is reduced. To achieve this harmony, the ISO must strive to present tangible, well articulated hazard observations and steer clear of any statement, feedback, or action that subverts the IC. To avoid trouble with the IC and to work toward a safe incident scene, the ISO should embrace the key concepts of authority, communications, and a solution-driven approach.

Authority   The assigned ISO must yield to the IC's authority and not pursue any argumentative approach to correcting tactics or strategies that the IC has implemented. The ISO who feels strongly about a request to change a situation has the responsibility to speak clearly, professionally, and rationally to the IC. Once the IC has shared his or her decision, the ISO must accept it and move forward. To belabor the point can be damaging to relationships and does not assist in resolving the concern. Perhaps the ISO can return to information gathering or offer peripheral solutions that can address the hazard in question. The IC–ISO relationship should be dynamic, just like the progression of the fire or incident. Perhaps the situation will change and the ISO's suggestion becomes valid to the IC.

Remember that the vision or outlook between the IC and the ISO is similar but different. The best expression of this concept is in a quotation from the U.S. Fire Administration's *Risk Management Practices in the Fire Service:*[3]

> *Incident Commander's View:* "Get the job done and operate safely"
> *Incident Safety Officer's View:* "Operate safely and still get the job done"

As already stated, the ISO has the authority of an incident commander only when an *imminent threat* is present. In all other cases, the ISO needs to work out issues in the framework of the IMS. That means all proposed changes and concerns go through the IC. When occasionally, the IC delegates the operations function to an operations section chief, the ISO may establish a link with the ops chief so that changes can be made at that level, and the IC should approve the communication link.

Communications A vast majority of multifirefighter fatality incidents fault communication failure, most of which fall into one of three categories: (1) not enough communication, (2) incomplete or fractured communication, and (3) a lack of communication prioritization (too much "chatter").

Using all the communications tools available, the ISO should maintain contact with the IC. The two primary means of communication typically used by the ISO are radio and face-to-face.

When using the radio to communicate, be mindful of the limitations and congestion that are often present. Although local procedures and policies are likely to guide your radio use, some radio use suggestions for the ISO are warranted.

- *Use the identifier "Safety" when speaking on the radio.* This is congruent with the IMS and has the potential impact of gaining attention. The use of day-to-day readiness identifiers for the ISO is not consistent with IMS practice. For example, by using his or her day-to-day designator of Battalion 101 when filling the ISO role at an incident, the ISO can cause confusion as to his or her role. At an incident, the best practice is to use *function* radio labels. For example:

  > *ISO:* Ventilation Group from Safety.
  >
  > *Ventilation Group:* Safety from Vent Group, go ahead.
  >
  > *ISO:* Vent Group from Safety, your secondary escape ladder is being moved to the bravo-charlie corner because of heat exposure. Acknowledge?
  >
  > *Ventilation Group:* Safety from Vent Group, we acknowledge secondary escape ladder is now at bravo-charlie.

- *Limit radio use to the communication of significant safety messages, hazards, and firm interventions.* Limiting your radio use accomplishes two things: First, it allows you to hear more of what everyone else is saying. Second, others who rarely hear from "Safety" are more apt to listen if and when you do break in.

While radio communication is often essential, the face-to-face method is most effective in communicating with the IC and outside crews. This method allows for dialogue and feedback, and it eliminates most of the barriers imposed by radio transmissions. Further, face-to-face allows both parties the opportunity to see if the message is understood and engage in spontaneous dialogue. As a rule, the ISO should have face-to-face communication with the IC every 15 minutes at routine incidents and more frequently if conditions or factors change. At prolonged incidents, the 15-minute face-to-face rule may not be practical or warranted. In these cases, the ISO and IC should agree on a face-to-face communication schedule.

■ Note

As a rule, the ISO should have face-to-face communication with the IC every 15 minutes at routine incidents and more frequently if conditions or factors change.

Solution-Driven Approach   For an incident commander, problem solving is not only required, but expected. Too often, however, subordinate officers bring problems to the IC to be solved. While the authority and responsibility to solve these problems rest with the IC, it seems reasonable to bring the problem *and* a solution to the incident commander for consideration. This *solution-driven approach* accomplishes two things: First, the IC gets a head start on problem solving—one solution is already drafted. Second, the IC can troubleshoot the solution and offer guidance to others or simply adjust the overall incident action plan accordingly (**Figure 11-7**).

ISOs should embrace the solution-driven approach, not only to present themselves as partners to the incident commander, but also to enhance their credibility. In many ways, ISOs are consultants to the IC in that they ask

Figure 11-7   *The ISO can be part of the solution path, as opposed to just bringing problems to the table.*

Figure 11-8 *The ISO should strive to be a consultant to the incident commander.*

questions, define problems and strengths, draft solutions, and offer recommendations to those who make decisions (**Figure 11-8**).

When a concern arises that requires attention (*not* an imminent threat), the ISO can use a standard solution-driven approach to communicate it. The approach can consist of four steps:

1. Here's what I see (a factual observation).
2. Here's what I think it means (your judgment about the hazard).
3. This is what I would do (your solution).
4. What do you think?

By ending with a question, you are acknowledging the IC's authority. This approach is very effective and can work in the most difficult of situations.

The National Incident "Typing" Scheme

NIMS integration center (NIC)
the center responsible for the ongoing development and refinement of various NIMS activities and programs

The lessons of recent disasters have motivated emergency response agencies to adopt typing systems for state and interstate mutual aid systems. The NIMS document has given responsibility to the **NIMS Integration Center (NIC)** to develop a national resource typing protocol. NIC is responsible for the ongoing development and refinement of various NIMS activities and programs. It is highly likely that NIC will use and expand the classic typing system that has been used by the wildland fire community.

The classic wildfire typing system has several quirks. It uses roman numerals (I, II, IV, etc.) and an inverted scale to differentiate capability (we'll use standard numerals here). Logic suggests that a Type 3 widget would have more capability than a Type 1 widget. This is not the case: Type ratings are ranked the opposite way, with lower numbers indicating higher capabilities.

Type classifications are given to resources like personnel, teams, facilities, supplies, as well as major equipment, like fire apparatus. Many street-oriented responders use typing to describe incidents. If firefighters say they are responding to a Type 3 incident, they are really saying that they are responding to an incident that is being managed by a Type 3 incident management team (IMT). An **incident management team (IMT)** is a trained overhead IMS team with specific expertise and are organized to deploy to incidents for management functions that exceed those available at the local level.

incident management team (IMT)

a trained overhead IMS team with specific expertise and organized to deploy to incidents for management functions that exceed those available at the local level

To be NIMS compliant, local fire departments will one day probably have to document the qualifications of their incident commanders (and ISOs) with regard to the nationally defined IMT typing scheme. NIC has already developed online NIMS training courses to help local fire departments pursue these qualifications.

The generally accepted definitions of IMT types are based on the level of expertise required to manage the expected situation and resources of an incident. IMT types and qualifications include the following:

- *Type 5: Local, agency, or jurisdiction specific:* Capable of functioning in an incident management system from the discovery of and the arrival at an incident up to and including a full operational period defined by the local jurisdiction.

- *Type 4: Multiagency or jurisdiction (automatic and mutual aid responses):* Qualified as a Type 5 and capable of functioning in an incident management system for multiple operational periods as assigned at major incidents.

- *Type 3: Regional:* Qualified as a Type 4 and capable of functioning in an incident management system that involves resources from multiple agencies and jurisdictions from local through federal levels.

- *Type 2: State:* Qualified as a Type 3 and capable of functioning in an incident management system that involves the utilization of significant numbers of state and federal level resources.

- *Type 1: National:* Qualified as a Type 2 and capable of functioning in an incident management system that involves the utilization of significant numbers of federal level resources.

Expanding the ISO Function Within the IMS

Despite the daunting list of functions in NFPA 1521, in most cases, the ISO can handle them. When the unusual, prolonged, or technically complex incident arises, the ISO can be overwhelmed and should request the assignment of an assistant safety officer (ASO). If the assignment of an ASO is still inadequate, the ISO may opt to request more personnel to help with all the safety functions.

Before explaining how the ISO role can expand within the IMS, let's identify some language issues related to incident command systems and the National Incident Management System (NIMS).

ICS/NIMS Language   Many fire departments still use Safety Sector when expanding the ISO role. Although Safety Sector has worked well for many, it presents language issues when applied to the NIMS. Defined simply, NIMS can be viewed as an umbrella concept, and ICS is the basic site-specific component of NIMS (as are multiagency coordination [MAC] and joint information center [JIC]). Specifically, ICS is the hands-on responder management part of NIMS. Some call this the field or area command component of NIMS. Even though NIMS is a national program designed for significant incidents, the federal government has tied fire service grant funding to NIMS compliance. At the local level, fire service leaders need to show that their local ICS can escalate and blend with NIMS when required. The official NIMS designator for the ISO working at an area incident command post is Safety Officer (SO). However, we will continue to use the title ISO to differentiate the position from an HSO.

Being NIMS-compliant should not be a big issue for the fire service; after all, NIMS and ICS both came from the fire service (they just vary in their use of language). Fire departments, other emergency response organizations, the NFPA, and similar services are working to show NIMS compliance for political and funding gains. At the same time, because the fire service leadership created ICS/NIMS, they have matured in the process and are therefore ahead of the game. Arguably, the fire service leadership can continue the maturity process by adjusting ICS and NIMS as they discover better ways of managing incidents. Recent catastrophic events prove this.[4] In this chapter, we show how the ISO function can expand using NIMS language and how the ISO function might mature in the future.

Local-level ISO Expansion   When a single ISO cannot perform all the safety functions at an incident, the ISO should request, from the IC, an ASO. Also, certain types of incidents (specifically, hazardous materials incidents (technician level), confined space operations, and trench rescues) mandate the appointment of an ISO with the appropriate training/certification.[5] When the designated ISO does not possess technician-level training or certification for the type of incident in question, appointing a technician-level–trained ASO as part of the ISO staff helps satisfy the safety needs of the technician-level teams. The technician-trained ASO can be titled assistant safety officer–hazmat (ASO-HM), or simply hazmat safety officer in hazmat incidents or assistant

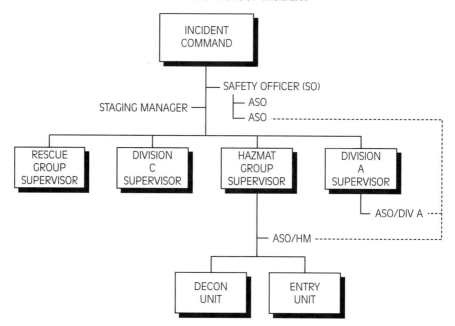

**THE USE OF ASSISTANT SAFETY OFFICERS (ASOs)
AT A DIVISION/GROUP INCIDENT**

Figure 11-9   *The use
of assistant safety
officers (ASOs) at a
division/group
incident.*

safety officer–rescue tech (ASO-RT) for a confined space incident. When
ASOs are utilized, the ISO should be located in the same area as the IC and
become a central contact point for the ASOs. This allows the ISO to manage
overall safety functions for all responders (including nontechnician-level responders) and all the concerns that shape the incident action plan.

Further ISO Expansion Options   Incidents that expand into a full-blown NIMS
event present significant challenges for the ISO. One or two ASOs may also
be overwhelmed. Currently, NIMS does not allow the safety officer to have
"units," as a plans or logistics function does. This area has room to mature.
To stay NIMS compliant, though, the ISO needs to remain a command staff
position, reporting to the incident commander while using multiple ASOs.

Another tenet of NIMS and ICS is to keep a manageable span of control.
Optimally, this is a ratio of five ASOs to the one ISO. If the ISO has ten assistants, how can they all be adequately managed? Once again, this is a weakness of the current NIMS document, which fails to give guided scalability for
the ISO function. In the Florida wildfires of 2005 and the gulf hurricane responses, the ISO sometimes had up to twenty-five assistants! Such situations
call for creativity to sidestep the NIMS weakness. If the ISO staff function requires more than five assistants, the ISO should consider breaking the safety

function into manageable parts. For example, all the ASOs out in the field monitoring firefighters can report to a single ASO, who reports to the ISO. ASOs working on other safety issues (such as injury report taking or food unit inspections) can similarly report to a more central ASO, who reports to the ISO. This approach can help the ISO stay NIMS compliant and still keep span-of-control issues manageable (**Figure 11-9**).

In the future, we may see the NIMS safety officer function become scalable in a more defined way (see A Solution to the NIMS Weakness: Scaling the ISO Function).

A Solution to the NIMS Weakness: Scaling the ISO Function

Large-scale incidents have demonstrated the need to integrate complex safety concerns into NIMS. In the past, well-meaning ISOs have had to develop ad hoc efforts to address and implement an integrated approach to safety, and such improvisation has uncovered many shortcomings. Following the WTC and Pentagon attacks, significant concern was expressed about the weakness of integrated safety efforts (NIOSH Publication 2004-144). ISOs assigned to Katrina, Rita, and the Florida wildfire disaster sites have expressed a similar desire to clarify how the ISO function should expand to meet the demands of the incident. It is not uncommon to have five, ten, or twenty ASOs assigned during a significant incident—which defies the span of control tenets of ICS.

Many working safety officers have recommended that safety management is more than a staff function; rather, it is a "scalable function." With NIMS compliance as a driving force, the progressive idea to create a *safety section* has been shot down, but the idea should be readdressed and pushed through NIMS and NFPA committees. Currently, NIMS does not address SO expansion other than to say assistants may be assigned. Given this limitation, let's outline a possible solution for the scalability of the ISO function.

First, the ISO shall communicate to the IC the need to expand the ISO staff position to ensure the coordination of safety management functions and issues across jurisdictions, functional agencies, and private-sector and nongovernmental organizations. The responsibilities assigned to the incident safety officer then can be expanded to help manage safety functions when the number of assistants, technical specialists, and safety stakeholders from multiple jurisdictions exceed a reasonable span of control for the ISO. The types of incidents that may require such expansion are:

- Incidents covering a large geographical area that include numerous branches, divisions, or groups.
- Incidents that involve several specialty teams (hazmat, tech rescue, and/or others.)
- Multiagency incidents in which a unified command is established.
- Type 1 and 2 incidents and those for which an area command is established.

(*continued*)

(*continued*)
- Incidents in which significant acute or chronic responder health concerns require coordination with local, state, federal, or other health representatives.

Current NIMS language allows for the generic title of assistant safety officer when additional personnel are assigned to assist the ISO. More specificity can help clarify the specific role these assistants bring to the safety function. These could include:

- *Deputy safety officer (DSO).* A member designated by the IC who, in the absence of the ISO, is delegated the authority to fill the ISO position
- *Assistant safety officer–hazmat technician(ASO-HT)*
- *Assistant safety officer–rescue technician (ASO-RT)*
- *Assistant safety officer–technical specialist (ASO-TS).* A person with unique or specific expertise that can support safety functions
- *Line safety officer (LSO).* An assistant who is trained as an incident safety officer and assigned to a branch, division, or group that is actively monitoring the safety issues for "hands-on" responders
- *Area command safety officer (ACSO)*

When expanding the safety staff, the ISO should be mindful of the span of control. It is suggested that safety functions be divided into units, which can include:

- *Line unit.* Line safety officers (LSOs) assigned to various branches, groups, and/or divisions
- *Acute and chronic health unit.* Personnel assigned to help research, develop, implement, and document responder health issues; also personnel assigned to inspect food and hygiene incident facilities
- *Technical safety unit.* Specialists with the training, certification, and/or expertise to help address responder safety issues
- *Area command unit*

Figure 11-10 shows how units can be integrated into the ISO function. Remember, scaling the SO function with *units* is not currently NIMS compliant; this is merely an attempt at creating dialogue to help address the NIMS shortcoming.

The ISO as Part of the National Response Plan

If the incident becomes national in scope, the ICS component (and ISO function) are likely to become an area command under NIMS. At the overhead or federal level, the National Response Plan is implemented and a joint field office (JFO) is established. The chief of staff at the JFO has an assigned safety coordinator to help the ICS safety officer. The safety coordinator is tasked with coordinating worker safety and health resources from various federal departments (Department of Defense, Department of Health and Human Services, Department of Labor/OSHA, among others). There is also a "Worker Safety and Health Support Annex" as part of the National Response Plan.[6]

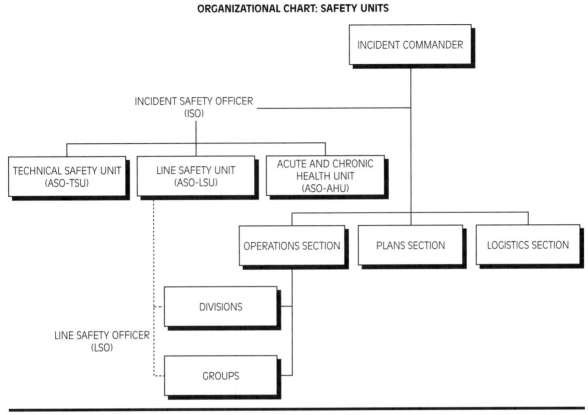

ORGANIZATIONAL CHART: SAFETY UNITS

Figure 11-10   *An organizational chart for scaling the ISO function into* units, *used to bridge an NIMS weakness.*

SUMMARY

Technically competent incident safety officers are only as effective as their ability to trigger safe behavior. These triggers can be classified as passive or active. Soft intervention (awareness, reminders, and suggestions) is the trigger most often used. ISOs need to avoid common traps that can render them ineffective, such as acting like a "bunker cop," getting sucked into performing hands-on tasks, and presenting a liability-protecting front. When working within command systems, ISOs understand and support

the authorities of an IC. Good communication practices and taking a solution-driven approach make for good IC relations. As an incidents grows in size and complexity, the qualifications required to manage safety functions also grows. A national typing scheme is being developed to help define these qualifications.

At times, a single ISO cannot address all the required functions. In these cases, the ISO should request ASOs from the IC. At large incidents, the scope of the ISO function may require assignees in numbers that exceed a reasonable span of control. In such cases, the ISO should create a system in which multiple ASOs report to a single ASO, who reports the ISO in order to be NIMS compliant. Using NIMS language and processes is warranted—

although recent incidents show that NIMS is not perfect. The lack of maturity and options in the NIMS model may one day lead to the creation of a scalable SO function to provide an integrated, incident-wide safety management approach.

KEY TERMS

Firm intervention An intervention to immediately stop, alter, or suspend an action or operation due to an imminent threat; more or less an official order to stop, alter, or suspend an action.

Imminent threat An activity, condition, or inaction that will most certainly lead to injury or death.

Incident management team (IMT) A trained overhead IMS team with specific expertise

and organized to deploy to incidents for management functions that exceed those available at the local level.

NIMS integration center (NIC) The center responsible for the ongoing development and refinement of various NIMS activities and programs.

Soft intervention An intervention to make crews, command staff, and general staff aware that a hazard or injury potential exists.

POINTS TO PONDER

The Alligators

Florida has seen its share of tourists, snowbirds, hurricanes, and wildfires. Florida fire departments have had to adapt to the challenges presented by each of the four, and, over time, firefighters and fire officers have learned to work especially well with each other across city, district, and county lines. The statewide response system in Florida has evolved into what is now considered one of the nation's best in responding to disaster. It was not always that way. When Florida started experiencing significant interface wildfires (1990s), area fire officers were burdened by the lack of training, equipment, and organizational procedure needed to deal with such large fires. Several chief officers reported that it was an amazingly difficult experience. Federal IMS teams and wildfire-trained crews responded to Florida in large numbers, bringing their significant experience gained from fighting wildfires in the dry, Western United States. Even with their experience, Florida wildfires presented interesting challenges that the outside teams never had to deal with: swamps, high ambient temperatures, and *alligators*.

(*continued*)

(*continued*)

During one wildfire, an Oregon fire crew experienced a close call with an alligator. Appropriately, the safety officer assigned to the fire decided to include alligator awareness training as part of the safety briefing for outside crews. Amazingly, that led to another problem. Armed with their new alligator information, outside crews became more comfortable in the alligator habitat. Word then got out that teams from other states were actually trying to catch an alligator and have their picture taken with it as a keepsake of their Florida experience. Reportedly, bets were placed on which team would get the first picture. Can you imagine the headache that presented for a safety officer? The SO was now facing serious safety *and* legal issues!

For Discussion:

1. Whom would you consult to develop safety briefings and training to address the alligator-related issues?
2. What legal issues could be involved with the capture of an alligator?
3. What unique issues in your jurisdiction would require special training or briefings for outside responders?

Note:

Thanks are due to the numerous Florida fire officers who shared their anecdotal experiences for this case study. Special thanks go to Assistant Chief Jeff Money, Brevard County Fire/Rescue, Florida. Any misrepresentations are the fault of the author.

REVIEW QUESTIONS

1. List two *passive* triggers that an ISO can employ to achieve safe behaviors.

2. List two *active* triggers that an ISO can employ to increase safety.

3. List the three ISO traps and discuss how each can render the ISO ineffective.

4. Who is ultimately responsible for incident safety?

5. Describe the organizational relationship of the ISO within the IMS.

6. What are the two primary communication tools the ISO uses?

7. As a rule, how often should the ISO communicate with the IC at routine incidents?

8. Define the proposed NIMS incident typing scheme for incident management teams.

9. Define how the ISO organizational structure can expand for local-level incidents.

10. Define how the ISO organizational structure can expand for large incidents.

ADDITIONAL RESOURCES

Brunacini, Alan V. *Fire Command,* 2nd ed. Quincy, MA: National Fire Protection Association, 2002.

Emery, Mark. "The 4-C Communication Model." *Health & Safety for the Emergency Service Personnel* 16, no. 9 (September 2005).

National Incident Management System. Washington, DC: U.S. Department of Homeland Security, March 2004. Available at: www.dhs.gov/interweb/assetlibrary.

National Response Plan. Washington, DC: U.S. Department of Homeland Security, last updated 2006. Available at: www.dhs.gov/interweb/assetlibrary.

National Incident Management System. Web site offering online training, ICS forms, graphics, and sample procedures: Available at: www.nimsonline.com.

"Safety Management in Disaster and Terrorism Response." *Protecting Emergency Responders,* Vol. 3. NIOSH Publication No. 2004-144. Available at: www.cdc.gov/niosh/docs/2004-144.

NOTES

1. Albert Schweitzer, *Great Quotes from Great Leaders* (Lombard, IL: Great Quotations, 1990).

2. Gene Chantler, "The Safety Officer: A Roundtable," *Fire Chief Magazine* (February 1993).

3. *Risk Management Practices for the Fire Service, FA-166* (Washington, DC: United States Fire Administration, 1996).

4. The challenges related to the World Trade Center collapse, Hurricanes Katrina, Rita, and Wilma, and the wildfire experience have shown that there is room for improvement in NIMS, ICS, and the ISO functions. (This statement is based on the author's observation from interviews, postincident investigations, and critiques). The creation of an integrated safety management system at large-scale incidents has been recommended by NIOSH in Publication No. 2004-144, "Safety Management in Disaster and Terrorism Response," *Protecting Emergency Responders, Volume 3,* Publication No. 2004-144 (Washington, DC: NIOSH, 2004).

5. Title 29 CFR 1910.120, *Hazardous Waste Operations & Emergency Response Solutions;* 29 CFR 1910.146, *Permit-Required Confined Spaces;* and 29 CFR 1910.1926, *Excavations, Trenching Operations* all list the requirement for an ISO with training/certification for the specific type of emergency operation.

6. The entire National Response Plan (NRP) and the NIMS document are available online at *www.dhs.gov/interweb/assetlibrary.*

Chapter

12

A BASIC APPROACH TO ISO DUTIES

Learning Objectives

Upon completion of this chapter, you should be able to:

- List two methods to achieve a systematic approach to ISO duties.
- List several advantages and disadvantages of using checklists, as well as four design considerations when creating them.
- Describe the differences between linear and cyclic thinking.
- List the four components of the ISO Action Model.
- Describe the four steps that help an ISO become integrated into an incident.
- List the three ISO general duties applicable to all incident types.
- List the four personal safety systems that the ISO needs to evaluate.

GETTING STARTED ISSUES

Without a systematic approach to accomplish all the ISO-related mental and physical tasks, the safety officer is destined to be mediocre. Experienced ISOs typically develop a uniform approach that is applicable to many types of incidents. Yet certain issues and frustrations challenge the ISO who is trying to develop the best approach to incident duties.

Perhaps the biggest issue (and frustration) facing the ISO is the prioritization of necessary physical and mental functions. Often, an ISO arrives at the incident scene after the initial fire attack or setup. After checking in with the incident commander, the ISO hopes to get a quick briefing on what has happened, what is planned, and what the IC needs. What's next? Often the IC requests a 360-degree scene survey; sometimes, the IC may have a specific question or concern (such as, "When is this building going to collapse?"). Other times, the IC leaves the options up to the ISO by saying, "I don't have the full picture here. Find out what's going on and report back." Worse yet, the IC may just assume that you know what needs to be done and gives no indication of priorities. It is not at all uncommon for an IC to have the ISO draft a quick action plan based on the current and predicted situation statuses and resources. Given all this, the ISO can find it hard to develop a starting place for the many items that need to be addressed—leading to frustration.

This chapter looks at several methods or approaches that help the ISO address all the required functions. In the early 1990s, approximately five hundred feedback/review sheets from the Fire Department Safety Officer's Association course, "Preparing the Fire Ground Safety Officer," were reviewed. In response to the question, "What can we add to the class to better address your needs?" respondents listed numerous frustrations and asked for tools to address them. Sample frustrations were as follows:

- There are no clear starting places for ISO duties.

- Existing ISO checklists are too short and/or too general to provide what is needed to be effective.

- The ISO must stay flexible and not be sidetracked with details that can obscure the "big picture." How do we do this?

- We need a strict and reliable method to take immediate action whenever the situation warrants.

- The expectation is that the ISO needs to "see all" and "know all" to be effective. We need a tool to help us accomplish this.

- Typical checklists imply that, once an item is checked off, it no longer needs attention—a dangerous assumption for an ISO. A checklist that includes everything that an ISO should address would be very long and unrealistic to apply to every incident.

• The ISO must be *reactive* to the needs of the incident commander as well as *proactive* in the prevention of injuries to firefighters. These opposing expectations create a priority clash at times.

This last point can be frustrating. The ISO role can be viewed as essentially both reactionary and proactive. On one hand, the ISO must look at what has already happened and offer solutions to correct unacceptable hazards or risk situations. On the other hand, the ISO has the opportunity to predict future events and make suggestions to minimize the effect of the events on the firefighter. Most incident commanders and working crews are receptive to the reactionary component; the situation is likely to be visual and explainable. Being proactive, however, is more subjective and places the ISO in a position to "sell" something that is not so obvious. In these cases, the ISO has to deal with opinions regarding the "likelihood" that the hazard in question will impact firefighters.

Addressing all the ISO functions and balancing reactive and proactive needs can be achieved by utilizing a systematic ISO approach to incidents. The two most common approaches to ISO incidents duties include the use of checklists and the use of action models. To be effective, checklists and action models must have certain qualities that make them usable. (See Qualities of Good ISO Checklists and Action Models.)

Qualities of Good ISO Checklists and Action Models

• **Flexibity:** Checklists and action models need to be flexible and adaptable to a multitude of incident types. Checklists may have to be developed to match certain types of incidents.

• **Cyclicity:** The ISO must be reminded to be cyclic in the use of checklists and action models. "Cyclic" refers to the concept that hazards, operations, and issues need to be revisited several times during an incident.

• **Proactive orientation:** The good ISO is always predicting or forecasting the next stage of an incident. The checklist or action model needs to reflect this need to be a step ahead.

• **Reactive orientation:** Having a process for the correction and communication of imminent or potential hazards or operations should be included in any system the ISO uses.

• **Archive friendliness:** Documenting observations, actions, and reminders can pay dividends when juggling multiple ISO functions. The checklist or action model used by an ISO should include some method or space for spontaneous note taking. The note-taking system should not be cumbersome and should be able to weather the incident environment (wetness, wind, cold, gloves, low light, and so on).

The remainder of this chapter details action models and checklists, then looks at an arrival process to integrate the ISO into an incident. We finish the chapter with a list of general ISO duties that are applicable at all incidents, regardless of type.

CHECKLISTS

One thing is certain in the fire service: *We love our checklists!* Virtually every arena of the fire service has adopted them because of all the benefits they provide. Some benefits are easy to see:

- They provide a quick reminder of things that need to be done.
- When you are distracted, they help you get back on track.
- They lend themselves to uniformity (from person to person doing the same task).
- Archiving is relatively simple.
- Changing the checklist is relatively simple within the framework of a fire department.
- Most formats are easy to understand.

It is easy to see why the fire service has embraced the checklist. The flip side of the coin is that checklists can hamper the ISO's effectiveness. Some disadvantages are:

- There is no one right way to perform the functions of the ISO.
- Checklists have a tendency to be either overly simple or amazingly complex.
- Once an item is checked off, the ISO may forget to revisit it.
- To cover a multitude of incident types, the ISO would have to carry a filing cabinet.
- Checklists imply an order for task completion, especially for the new or inexperienced ISO.
- They may be subject to subpoena in legal matters.

Many templates and sample forms are available to help create a checklist designed to be used in a systematic approach to ISO functions. Even when a jurisdiction (especially at large, multiagency incidents) requires pre-designed checklists and forms, the ISO can easily develop a helper checklist or note pad that makes up for the inadequacies of defined forms. When *required* checklists and forms are not an issue, the ISO is encouraged to create a checklist that takes into account local variables. Some thought and considerations should be given to the design and content of the checklist (see Design Considerations for Checklists). Appendix C of this book includes several checklists that have been developed by working ISOs and may be helpful when designing your own.

Design Considerations for Checklists

- Simple-column format
- Easy to read in low light
- Room for notes and diagrams with space to accommodate grease pencils or water-resistant markers
- Easy to differentiate from other similar checklists (for example, use different-colored templates or big bold titles)
- A reminder area for required postincident actions

ACTION MODELS

action model
a template that outlines a mental or physical process to be followed

An **action model** is basically a template that outlines a mental or physical process to be followed. Some ISOs prefer the flexibility and adaptability of action models as opposed to checklists. The biggest advantage of action models is that they furnish a template in which to process multiple inputs.

The first edition of this book presented an action model developed by the author and Deputy Chief Terry Vavra of the Lisle-Woodridge Fire District (Illinois). The model arose from the frustrations experienced with designing checklists for the ISO. We believed that a simple, easy-to-apply model could be created that was adaptable to most incidents, overcoming those frustrations. One of the key ingredients in designing the action model was the need to remind ISOs to be "cyclic" in their thinking and to stay open to changing inputs.

■ Note
The biggest advantage of action models is that they furnish a template in which to process multiple inputs.

Cyclic Thinking

All of us use a linear thinking process to handle incidents, that is, a process having a defined starting point and a desired end point. During the thinking process, inputs are made along the linear path. At some point the person making decisions may reach his or her maximum capcity for input, causing stress. Combine this point of maximum input with the focus to reach an ending point, and it is easy to see that some hazards may be overlooked. This point was highlighted following the tragic Storm King Mountain fatalities during the South Canyon Fire in Colorado (see Overload at the South Canyon, Colorado, Fire).

Overload at the South Canyon, Colorado, Fire

Fourteen firefighters died in a sudden blowup of the South Canyon Fire in 1994. Following the incident and subsequent investigations, human error and communication breakdown were cited as contributors to the incident. Ted Putnam, PhD, specialist for the Forest Service's Missoula Technology and Development Center, wrote an article entitled "Collapse of Decision Making," which appears in the publication, *Findings from the Wildland Firefighters Workshop*[1] and discusses the psychological elements that lead to failed leadership during wildland incidents. In the article, Dr. Putnam makes a case for the traps of linear thinking by discussing decision-making models and conclusions based on numerous studies. Dr. Putnam concludes that increasing incident stress actually leads the decision maker to minimize the number of inputs being considered, and the person regresses toward a more habituated behavior. Checklists can contribute to linear thinking.

Safety
It is imperative for ISOs to create an environment in which they can stay open to multiple inputs and maintain a high degree of situation scanning and awareness, that is, maintain cyclic thinking.

Note
A system of cyclic, or recurring, evaluation by the ISO can help eliminate the trap of underestimating hazards.

ISO action model
a cyclic, four-arena model that allows the incident safety officer to mentally process the surveying and monitoring functions of typical incident activities and concerns

The IC must be a linear thinker: Establish a path and work toward a positive conclusion. It is imperative for ISOs to create an environment in which they can stay open to multiple inputs and maintain a high degree of situation scanning and awareness, that is, maintain cyclic thinking. In many ways, the ISO is the "what-if" thinker who tends to overestimate hazards. With this in mind, the ISO action model was developed using a circular image—a reminder to stay open to the hazards that may eventually cause injury to firefighters. A system of cyclic, or recurring, evaluation by the ISO can help eliminate the trap of underestimating hazards.

The ISO Action Model

The **ISO Action Model** is a cyclic, four-arena model that allows the incident safety officer to mentally process the surveying and monitoring of typical incident activities and concerns. As shown in **Figure 12-1**, the model calls for the ISO to address four general arenas. Neither a starting place for the model nor a direction of flow should be inferred. Upon the ISO's check-in with the incident commander, a starting place may be assigned. If this is the case, the ISO simply jumps into the cycle as directed. If the IC does not select a starting place, the ISO can start where he or she feels attention is warranted. Once on the cycle, the ISO should conduct an initial survey of each arena, and then monitor the applicable concerns in each. In essence, the ISO performs a mental evaluation of the conditions, activities, operations, or probabilities in each arena.

The four Action Model components requiring evaluation and attention can be thought of as the *Four Rs.*

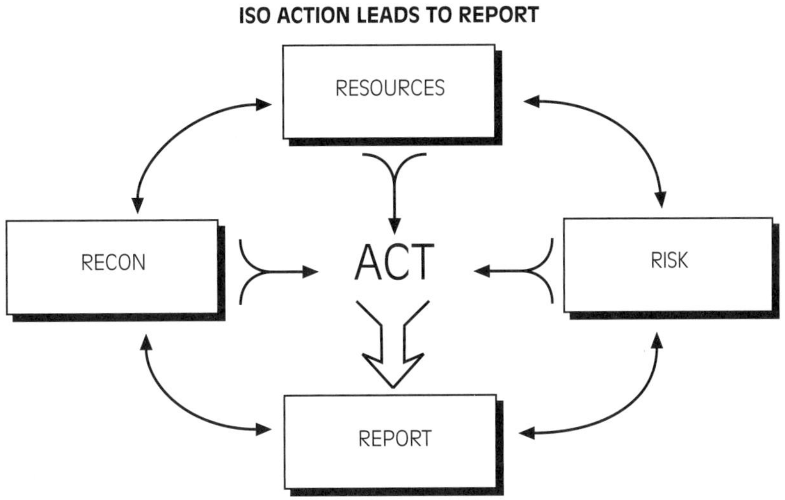

ISO ACTION LEADS TO REPORT

Figure 12-1 *Any of the Action Model components can cause the ISO to take action. Any action, however, leads the ISO to report to the incident commander.*

Resources   It is easy to say that there are *never* enough people or equipment to handle a significant incident. Incident commanders earn their respect by their ability to manage available resources. Similarly, the ISO must make an evaluation of the resources on hand and determine whether they match or support the action plan. Specifically, ISOs need to evaluate the resources of *time, personnel,* and *equipment.* They should use the skill of "reading firefighters" when evaluating resources. If any of the resources are thin, the incident action plan does not match the resources and intervention is warranted.

Reconnaissance   Most incident commanders agree that having an incident safety officer available to help with a 360-degree scene survey is essential to improving scene safety. This survey is often called the **reconnaissance, or "*recon*,"** an exploratory examination of the incident scene environment and operations. The effective ISO uses this recon trip to read smoke, read the building, and read hazardous energy.

reconnaissance (recon) an exploratory examination of the incident scene environment and operations

Risk   Chapter 8 (Reading Risk) discussed the process of deciding whether a situation or an operation presented an *acceptable* risk. On-scene, the incident commander must put the concept of reading risk into practice through the observations and efforts of the incident safety officer.

Report   What seems like an obvious responsibility of the incident safety officer is actually one of the most often forgotten. Many of the conflicts between incident commanders and incident safety officers can be resolved through timely, appropriate communications (see Chapter 11). Additionally, the report phase of the Action Model reminds the ISO that written reports, safety briefings, and review of incident action plans are necessary.

The word "act" on the action model in **Figure 12-1** refers to any ISO intervention. When an intervention is made, the ISO is reminded to report this to the IC.

By addressing each of the Four Rs continuously and cyclically, ISOs apply a systematic approach to their duties. Further, the model can help reduce what could, at times, be an overwhelming incident for ISOs.

THE ISO ARRIVAL PROCESS

Whether you choose a checklist or an action model, you still need a process to become integrated into an ongoing incident. Rarely does a predesignated duty ISO arrive first on-scene. If the duty ISO does arrive first, it makes sense for the ISO to take on the role as an initial incident commander. More often than not, a first-due company arrives, gives a status report, establishes command, and declares an initial action plan or mode of operation. If the duty ISO arrives after the first-due company but before a designated command officer, the ISO should probably assume command after connecting with the first-due (working) company officer. Granted, this action is subject to local policy but is consistent with model IC systems. In most cases, the ISO arrives (or is appointed) after the establishment of a command post and stationary IC. When the duty ISO does arrive (or the ISO assignment is given), a process of integration into the incident should be mandated to ensure that the ISO is dialed into the situation. The following steps can help ISOs get dialed in:

1. *Confirm the ISO assignment.* Upon arrival at an incident, a predesignated duty ISO should meet with the incident commander to confirm that the IC wants and/or needs the ISO position filled. Some fire departments have a policy that the IC does not have a choice—assigning an ISO is a must. Occasionally, the IC may need you to take another command staff assignment or act as a group or division supervisor. In these cases, the IC retains safety functions.

2. *Collect information.* Upon confirmation, the ISO needs to gather information. It is hoped that an IC would brief the ISO on the overall incident action plan (IAP). In addition to the IAP, the ISO should inquire about the status of the situation and resources. This is also a good opportunity to ask about known hazards, the establishment of control zones, and the status of rapid intervention crews/companies and a rehab system.

3. *Confirm communication links.* Take the time to confirm assigned tactical radio channels and, if policy does not address the issue, ask the IC which radio frequency he or she prefers to be used. Additionally, offer a face-to-face communication schedule so that the IC knows what to expect from you. It may be desirable for the IC to announce over the tactical radio channel that "safety" has been assigned. (For example, "All fireground personnel from Fourth Street Command, Battalion 101 is now "Safety.")

4. *Don appropriate identification and PPE.* The ISO should don PPE that is appropriate for the likely potential hazards, as well as an ISO-identifying vest or helmet, and check into the personal accountability system. From here, the ISO most likely performs a recon of the incident and begins the ISO duties using a systematic approach.

GENERAL ISO DUTIES

Regardless of the type of incident, several expectations are understood when the IC has delegated the safety task to an ISO. NFPA 1521 lists these as incident safety officer functions. The standard goes on to divide functions into "incident scene safety" (applicable at all incidents) and specific functions at various types of incidents (fire suppression, hazmat, and so on). Later chapters of this book look at specific functions at various types of incidents. Here, the incident scene safety functions are presented as general duties applicable to all types of incidents. As with the ISO Action Model, no order is implied in accomplishing these duties.

! **Safety**
When incidents are judged or determined to be recovery in nature, the ISO should suggest strategies to reduce risk taking to the IC.

Monitor the Incident

The ISO shall monitor the incident action plan, conditions, activities, and operations to determine whether they fall within the criteria as defined in the fire department's risk management plan. With this monitoring comes the need to report the status of conditions, hazards, and risks to the incident commander. When the perceived risks are not within the fire department's risk management criteria, the ISO should take action. For example, when incidents are judged or determined to be recovery in nature, the ISO should suggest strategies to reduce risk taking to the IC.

! **Safety**
Failure to adjust the incident action plan to changing conditions has been cited as a contributing factor in numerous firefighter fatality investigations.

In addition to monitoring risks, the ISO should make a judgment, by means of repeated recon, whether the operating crews are effecting change as intended by the incident action plan. It is incumbent on the ISO to communicate to the IC if operating crews are not achieving the desired outcomes for an incident action plan. Note: Failure to adjust the incident action plan to changing conditions has been cited as a contributing factor in numerous firefighter fatality investigations. Communicating to the IC that the desired results are not being achieved or that incident conditions are deteriorating too fast for positive results to be achieved should lead to a change in the action plan. If the IC does not change the action plan, the ISO should evaluate whether an imminent threat exists and process a firm intervention.

Address Personnel Safety Systems

Many protective systems have been developed to help fire departments increase the safety of responders at an incident. The ISO's role is to evaluate the systems and take steps to make sure they are working as planned. Here's a list of the more important systems that the ISO should evaluate:

Personnel Accountability Systems   Check to see that the fire department's personnel accountability system is being utilized effectively. The effective use of an accountability system does not end with a check on the passport status board. You must also evaluate crews to make sure they are working within the framework of the action plan. Failure to work within the framework of the action plan is called **freelancing.** In the fire service, freelancing is viewed as a dangerous and deadly enterprise. Firefighters have been killed and seriously injured while engaged in a freelance operation, that is, an operation or task being performed unknown to the incident commander or other working crews. Freelancing is most often attributed to a lone worker, although the term can be equally applied to a rogue crew that has determined that a task needs its personal attention. Either scenario is potentially dangerous.

The solo firefighter is probably the deadlier of the two scenarios because of the what-if potential (**Figure 12-2**). What if the firefighter falls through the floor? What if the firefighter suffers a heart attack? What if the firefighter gets lost? What if the firefighter pushes the fire back on other crews? The what-ifs are endless and lead to one conclusion: Nobody will know that the firefighter experienced an emergency or injury. This is all in addition to the possibility that lone firefighters may get sucked into a situation that requires skills they may not possess or that may require more than one person to accomplish the task.

ISOs should keep a close eye on working crews and develop an eye for catching lone workers. They can also apply their basic knowledge of fireground operations and predict situations that lead to freelancing. For example, a crew is performing an assigned task and they need additional equipment. Often, the crew breaks up and one member (a gofer) is sent for the missing tool. The incident turns tragic as the gofer becomes lost, is trapped, or is distracted by a more demanding need. Likewise, the person waiting for the gofer to return may look to find ways to be productive, leading to more what-if hazards. As another example, a firefighter, especially an inexperienced one, who is assigned to a seemingly "trivial" task, such as monitoring

freelancing
failure to work within the framework of an incident action plan

Figure 12-2 *The solo firefighter presents the worst and potentially most dangerous form of freelancing.*

an exposure line or a positive pressure fan, may search for any excuse to get where the action is. Similarly, members of a crew or team may have completed a task but do not want to rotate to staging; they want to stay involved so they look for other tasks that need to be addressed in the action area. This is well intentioned but outside the action plan. Unfortunately, the incident commander may or may not be aware of the need for the crew's self-assigned task and may have higher priorities in mind for the crew. In these cases, you can catch the freelancing only if you have an intimate knowledge of the incident action plan.

Need for Control Zones   Offer judgment to the incident commander on the establishment of collapse zones, IDLH zones, and/or no-entry zones. Ensure that established zones are communicated to all members present on-scene. Whenever possible, control zones should be identified with colored hazard tape, signage, cones, flashing beacons, fences, or other appropriate devices (more information on establishing control zones can be found in Chapter 13).

Regarding hazard tape, the ISO should use a distinctly identifiable tape when marking responder no-entry zones. Yellow barrier (caution) tape is often used to identify hazards for the general public, and firefighters routinely duck under it. Predesignating a unique tape as a responder exclusionary barrier sends the message that even firefighters must stay out.

Radio Transmissions   It takes a keen ear to listen to radio chatter and spot transmission barriers that could result in missed, unclear, or incomplete communications.

Rehab Effectiveness   Ensure that a rehabilitation component has been established and is addressing rest, hydration, active cooling, medical monitoring and care, energy nutrition (food and electrolyte replacement), and extreme conditions. This is not to say ISOs set up and run the rehab component; they just make sure it is functioning and effective (**Figure 12-3**). They also

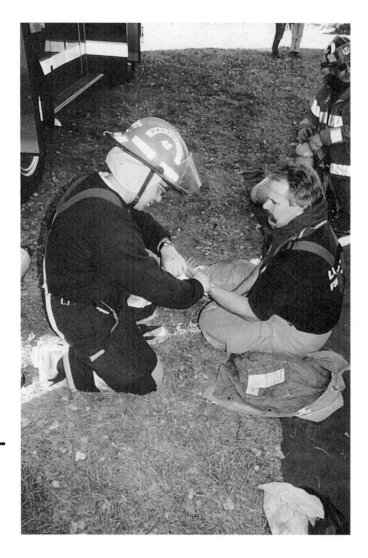

Figure 12-3 *Rehab efforts should also include provisions for quick medical checks of working firefighters.*

evaluate the need to activate critical incident stress procedures in response to the size, scope, or potential emotional impact of the incident. If warranted, they should make the request to the IC.

Define Other Needs

Other needs can be defined as the miscellaneous issues that ISOs should be addressing at all incidents. While not all-inclusive, the following short list includes items that are typical at most incidents.

- Evaluate motor vehicle scene traffic hazards and apparatus placement, and take appropriate actions to mitigate hazards. The number of firefighters struck by traffic at incidents is alarming. Focus extra effort on this area.
- Survey and evaluate hazards associated with the designation of a landing zone and interface with helicopters (**Figure 12-4**).
- Communicate to the incident commander the need for assistant incident safety officers due to the need, size, complexity, or duration of the incident.

These general duties apply at all types of incidents. In the next few chapters, we single out specific incident types and look at ISO functions that address the unique features of each.

Figure 12-4 *The ISO should evaluate the helicopter LZ to ensure that personnel will not be endangered if a mishap were to occur.*

SUMMARY

The ISO assignment can be overwhelming due to the multitude of issues that an ISO must address. The expansive and dynamic nature of the job can create frustrations without a systematic approach to the assignment. The two most common systematic approaches to the ISO function make use of checklists and action models. Checklists are probably used more often due to their numerous advantages. The ISO, however, understands the limitations of checklists and may prefer an action model. Action models are designed to *process* observations, issues, and actions. The use of the ISO Action Model presented in this chapter reminds ISOs to be cyclic in their thinking (not linear). Linear thinking is essential to incident handling but has the tendency to underestimate hazards. The ISO Action Model highlights four component areas that need to be processed: resources, reconnaissance, risk, and report. Any of the four components might trigger action by the ISO.

Upon arrival at an incident, ISOs should employ a process to get "dialed in." To do so, they should confirm the ISO assignment with the IC, collect incident information (such as the incident action plan), confirm communication links, and don appropriate PPE and position identification. After checking in, ISOs use a systematic approach to perform their duties. NFPA has outlined general ISO functions that need to be accomplished at all incidents. These functions can be grouped into three actions: monitoring risks and operational outcomes; addressing personal safety systems, like control zones, radio transmissions, and rehab; and defining other needs like traffic issues and additional ISO assistance.

KEY TERMS

Action model   A template that outlines a mental or physical process to be followed.

Freelancing   Failure to work within the framework of an incident action plan.

ISO action model   A cyclic, four-arena model that allows the incident safety officer to men-

tally process the surveying and monitoring functions of typical incident activities and concerns.

Reconnaissance (recon)   An exploratory examination of the incident scene environment and operations.

POINTS TO PONDER

The New ISO

Captain Franks was recently reassigned from his truck company officer position to that of a dedicated shift safety officer. His department has only recently adopted a duty shift safety officer program, and the department chose Franks and two other captains to attend an ISO training academy at another department to learn their new assignment. Captain Franks enjoyed the academy and felt like a

(*Continued*)

(*Continued*)

dedicated ISO position could make a difference. After a few weeks of learning his nonincident (HSO) responsibilities, Franks began longing for an opportunity to practice his newly acquired ISO skills. His wishful thinking was soon realized when his shift was dispatched to a working fire in a commercial building. Upon arrival at the incident, Franks was overcome with a feeling that the incident was not going well: lots of radio chatter, firefighters rushing around, thick black smoke coming from the second floor of a large storage building, and lots of vehicle traffic congestion. Franks dressed in his PPE, grabbed his radio, clipboard, and safety officer vest, and proceeded to check in with his battalion chief, who had assumed command. While walking to the IC, Franks looked at the first item on his ISO checklist, which reminded him to check in with the IC and ascertain the incident action plan. The IC briefly acknowledged Captain Franks' arrival and told him, "We have a good fire ripping on the second floor and all hell is breaking loose." The IC then rapidly listed a multitude of items that he needed Franks to accomplish: Do a 360, check the status of the attack crew who haven't been heard from, do something about the traffic congestion, find out why the truck company hasn't gotten the roof opened, and give a read on collapse potential. Captain Franks felt slightly overwhelmed, but took a deep breath and set off to perform his recon.

For Discussion:

1. What can Captain Franks do to better manage his overwhelmed feeling?
2. What information would you like to obtain before addressing the IC's needs?
3. Compare and contrast the use of checklist and action models for an incident that is seemingly chaotic.
4. What types of freelancing situations can you expect to arise at this incident?

REVIEW QUESTIONS

1. What two methods help the ISO achieve a systematic approach in addressing ISO duties?

2. List several advantages and disadvantages of using checklists.

3. What considerations should be considered in the design of a checklist?

4. What is one of the biggest traps of linear thinking?

5. List the four components of the ISO Action Model.

6. Describe the four steps that help an ISO become integrated at an incident.

7. List the three ISO general duties that are applicable to all incident types.

8. What four personal safety systems need to be evaluated by the ISO?

9. What are the two forms of freelancing?

10. How should a responder exclusionary zone be marked?

11. An effective rehab component should include what elements?

12. List three "other needs" that should be considered by the ISO.

NOTES

1. T. Putnam, "Collapse of Decision Making," *Findings from the Wildland Firefighters Workshop* (Missoula, MT: USDA Forest Service Publication 9551-2855 MTDC, July 1996).

Chapter 13

THE ISO AT STRUCTURE FIRES

Learning Objectives

Upon completion of this chapter, you should be able to:

- Discuss the relationship of risk-taking to incident benchmarks.
- With respect to structure fires, list the two factors that can help judge operational effectiveness and the three resource considerations.
- Name the three communication ingredients to an effective PAR.
- Define zoning strategies.
- List four examples where an ISO should request ASO assistance at structure fires.
- Describe what is meant by "rescue-profile."
- List the three dimensions that need to be defined during environmental reconnaissance.
- Name the five crew-exposure considerations.
- List several unique hazards at strip-mall structure fires.
- List four ISO functions and six ASO functions at high-rise fires.

INTRODUCTION TO SPECIFIC INCIDENT TYPES

Starting with this chapter, we look at the incident safety officer's role at specific types of incidents. Topically, we will devote a chapter to structure fires, hazmat incidents, technical rescues, and wildland/interface fires. Each of these chapters has a standard format to address the issues specific to the incident type. The format begins with general ISO duties, progresses to issues associated with the *Four Rs* of the ISO Action Model, and end with specific concerns unique to the incident type. This approach is intended to create a sort of "desk reference" and serve as tool to be utilized for ISO proficiency training.

ISO GENERAL DUTIES AT STRUCTURE FIRES

Of all the incident types to which fire and rescue departments respond, structure fires can be considered the most risky for numerous reasons, the greatest of which is the compressed time window that a fire department has to make a difference. Growth-stage fires in a building can change in minutes, if not seconds, and ISOs must rapidly read structure fires (smoke, building, and risk). Starting with the general ISO duties (from Chapter 12), we explore the issues associated with structure fires.

Monitoring Issues at Structure Fires

ISOs need to monitor two general areas at structure fires: (1) risk and (2) operational effectiveness.

Risk   The risks associated with structure fires are usually tied to tactical priorities and incident benchmarks (**Figure 13-1**). The ISO looks at the tactical priorities and determines if the risks being taken match the department's preestablished risk-taking criteria. Obviously, firefighters engaged in active victim rescue operations may have to face significant risk to their lives. Conversely, risk taking should be reduced drastically during overhaul. While such differences may seem obvious, injury and deaths statistics suggest that fire fighters do not always practice the obvious. The ISO needs to serve as the "risk cop" to make sure responders are following established risk guidelines. We get more specific about risk taking in the ISO Action Model risk component later in this chapter.

Operational Effectiveness   The ISO needs to make a judgment that the action plan is making progress toward achieving the objectives of the incident action plan (IAP). If the desired results are not being achieved, or if incident

TACTICAL PRIORITY	BENCHMARK	RISK-LEVEL
SEARCH AND RESCUE	"ALL CLEAR"	RISK A LIFE TO SAVE A KNOWN LIFE
FIRE CONTROL	"UNDER CONTROL"	TAKE A CALCULATED AND REDUCED RISK TO SAVE VALUED PROPERTY
PROPERTY CONSERVATION	"LOSS STOPPED"	TAKE NO RISK TO SAVE WHAT IS LOST

Figure 13-1 *In structure fires, benchmarks and acceptable risk levels are associated with tactical priorities.*

conditions are deteriorating faster than positive results can be achieved, the ISO should inform the IC. At most fires, however, ISOs are outside the building, performing recon. How can they tell whether the crews are being effective? They must utilize their developed skills for reading smoke and reading buildings to make judgments about operational effectiveness. For most fires, a positive outcome is likely when the fire is adequately ventilated (smoke pressure is relieved from the building) and fire flow is met (GPM quenching).

The ISO should observe signs that a fire attack is being successful. When no progress is evident, the ISO should determine whether the attack is failing due to inappropriate fire flow to the fire seat or due to the lack of adequate ventilation. Failure to make this evaluation may lead to a situation in which firefighters are overrun by the fire. Rapidly expanding steam that overtakes issuing smoke is a good indication that fire flow is being met. Pockets of steam that pale in comparison to the velocity, density, and color of smoke generated may mean that the fire is releasing more heat than the stream can match. In these cases, the ISO should be quick to communicate to the IC that fire flow and ventilation are inadequate. More often than not, inadequate ventilation efforts impede fire control.

PAR (personnel accountability report)
an organized reporting activity designed to account for all personnel working an incident; to be truly effective, PAR radio transmissions should include assignment, location, and number of people in the assignment

Personal Safety System Issues at Structure Fires

Accountability Systems When checking into the passport accountability system, the ISO should ask the accountability manager if things are going well. The reply can help the ISO evaluate whether the system is tracking firefighters effectively and the system is in harmony with the IAP. If not, the ISO may consider requesting that the IC calls for a **PAR** (personnel accountability report). The PAR is an organized reporting activity (usually via radio communication) designed to account for all personnel working an incident. To be truly effective, a PAR should include radio transmissions that include the assignment, location, and number of people in the assignment (see Effective PAR Communications).

Effective PAR Communications

Example of _Ineffective_ PAR Communication

Fire Attack Team 1 from Accountability: Call for PARs.

Accountability from Fire Attack 1: We have PARs with three.

Fire Attack 1 from Accountability: Copy PARs of three.

Example of _Effective_ PAR Communication:

Fire Attack Team 1 from Accountability: Call for PARs.

Accountability from Fire Attack 1: We have PARs with three, performing fire control on the second floor, bravo side.

Fire Attack 1 from Accountability: Copy PARs of three, performing fire control on the second floor, bravo side.

---

hot zone
the area presenting the greatest risk to members and often classified as an IDLH atmosphere; denote hot zones with red tape

warm zone
a limited access area for members directly aiding or supporting operations in the hot zone; denote warm zones with yellow tape

cold zone
establishes the public exclusion or clean zone; there are minimal risks for human injury and/or exposure in this zone; denote cold zones with green tape

PARs should be accomplished periodically at structure fires. Some fire departments mandate PARs every 15 minutes while firefighters are working in an IDLH environment or other high-risk environments. Additionally, some fire departments outline certain situations or changes that should trigger a PAR, for example:

- Anytime the operational mode has changed (i.e., switch from offensive to defensive)
- Anytime an incident benchmark has been achieved (all clear, fire under control, etc.)
- Following the report or witnessing of a flashover or collapse
- After the report of missing or trapped firefighters (Mayday report)

Control Zones   The ISO is responsible for establishing control zones (or adjusting established zones) at fires. The 2008 edition of NFPA 1521 (appendix) defines the traditional hot, warm, and cold zones, then goes on to add an important fourth: the _no-entry zone_. Additionally, the appendix suggests the color of barrier tape that should be used to discriminate the zones.

- _Hot zone:_ The area presenting the greatest risk to members and often classified as an IDLH atmosphere. Denote hot zones with red tape.
- _Warm zone:_ A limited access area for members directly aiding or in support of operations in the hot zone. Denote warm zones with yellow tape.
- _Cold zone:_ Establishes the public exclusion or clean zone. There are minimal risks for human injury and/or exposure in this zone. Denote cold zones with green tape.

No-entry zone
areas where no person—including firefighters, police, other responders, or the general public—should enter due to the serious or unpredictable nature of a hazard or condition; denote no-entry zones with red/white chevron tape

IDLH zone
areas in or around the building where working firefighters are exposed or may become exposed to smoke and heat; persons working in an IDLH zone shall have a partner, work under the two-in/two-out rule, and be tracked through an accountability system

collapse zone
areas that are exposed to trauma, debris, and/or thrust of a collapse; a more specific form of a no-entry zone "

support zone
areas where firefighters, other responders, IMS staff, and apparatus are operating or staged; the general public should *not* be allowed to wander into the support zone

- *No-entry zone:* Areas where no person—including firefighters, police, other responders, or the general public—should enter due to the serious or unpredictable nature of a hazard or condition. Denote no-entry zones with red/white chevron tape.

The NFPA control zone suggestions are a good starting place, although many ISOs prefer to add clarity to the general definitions. Defining zones as IDLH, no-entry, collapse, and support is clearer and minimizes miscommunication (see A Better Way to Label Zones).

A Better Way to Label Zones

IDLH zone: Areas within or around the building where working firefighters are exposed or may become exposed to smoke and heat. Persons working in an IDLH zone shall have a partner, work under the *two-in/two-out* rule, and be tracked through an accountability system.

No-entry zone: Areas where no person should enter due to the serious or unpredictable nature of a hazard or condition. "No person" means *nobody*—including firefighters, police, other responders, or the general public. Examples include, but are not limited to, damaged electrical equipment areas, unstable earth or geographical hazards, collapse zones, and the like.

Collapse zone: Areas that are exposed to trauma, debris, and/or thrust of a collapse. A collapse zone is a more-specific form of a "no-entry zone."

Support zone: Areas where firefighters, other responders, IMS staff, and apparatus are operating or staged. The public should *not* be allowed to wander in the support zone, although escorted media representatives might be allowed at the discretion of the IC and/or PIO.

Radio Transmissions   The fireground is filled with significant noise, an obvious barrier to effective radio communications that leads to missed, incomplete, or confusing communications. Additionally, crews are performing laborious tasks that require dexterity and concentration, another barrier to radio transmissions. Even so, the ISO needs to listen for unanswered radio calls and make a judgment whether the unanswered call indicates communication barriers or the need for rapid intervention to locate the crew not responding. Radio message priority is also important at the structure fire. Crews engaged in IDLH environments should have priority to report conditions, needs, and progress. The ISO needs to listen to radio traffic and, when the sense of priority is not congruent with the risks being taken, intervene if necessary. When listening, pay attention for trigger words or phrases that indicate a developing problem, for example:

- Mayday
- Urgent

■ **Note**

Defining zones as IDLH, no-entry, collapse, and support is clearer and minimizes miscommunication (see A Better Way to Label Zones).

Safety

Crews engaged in IDLH environments should have radio priority to report conditions, needs, and progress.

- Emergency traffic
- Unintelligible yelling
- I'm lost
- Look out!

Rehab   To evaluate rehab effectiveness at structure fires, the ISO needs to focus primarily on the effects of heat, physical exertion, and weather exposure. Firefighters have been known to disregard these effects if there is still "fun" things left to accomplish. Look for signs that the rehab effort is not being effective (**Figure 13-2**). Firefighters leaving the building for a bottle change likely need active core temperature cooling, hydration, and, food (if they have not eaten in more than a few hours). Department policies should

Figure 13-2   *Firefighters overdue for rehab are at high risk for injury.*

dictate a mandatory rehab cycle for firefighters working at structure fires. When this is not directed by policy, the ISO should encourage mandatory rehab after every air cylinder use or equivalent work period.

While interior crews typically receive the most rehab effort, the ISO cannot forget to check up on the outside responders. We rarely consider the rehab needs of apparatus operators, rapid intervention crews and companies, and command staff and assistants when fighting structure fires. If the outside weather is extreme, responders exposed to the elements also need relief from heat or cold and dehydration.

Defining Other Needs at Structure Fires

Traffic For the most part, the traffic around structure fires comes under control quickly, as law enforcement officers block streets where hose lines are laid. The "invasion" of large apparatus at the structure fires also helps control traffic issues. The greatest traffic risk to firefighters at structure fires is when they are arriving or moving apparatus, especially when water-tender shuttle operations are underway. For shuttle operations, the ISO should see that a traffic flow plan is communicated to all responders. The ISO should also evaluate apparatus placement, ongoing apparatus movement, and traffic lanes to see if hazards exist. Using the soft intervention of "awareness" is likely to be the best method to address such situations. Remind apparatus operators to slow down, use spotters (in congested areas or when backing up), and expect the worst. Remind firefighters to "look both ways" and never to run or jog around apparatus movement areas.

● **Caution**
The greatest traffic risk to firefighters at structure fires is when they are arriving or moving apparatus, especially when water-tender shuttle operations are underway.

Need for ISO Assistance In certain situations, the ISO should request an assistant safety officer(s) at structure fires:

- *Large buildings with significant fire involvement:* You are likely to need help monitoring the building, fire conditions, crew effectiveness, and/or apparatus issues.
- *When a "plans section" is established at the fire:* The ISO needs to provide input on action plan changes and strategic-level issues. ASOs are needed to accomplish the ISO "field component" (recon, rehab evaluations, and similar operations).
- *Fires in buildings with unusual or unique hazards:* The ISO may want an ASO to handle the typical ISO functions while the ISO meets with building representatives or technical specialists.
- *Anytime the ISO is requested to go into an IDLH environment:* In these cases, to retain an outside viewpoint, the ISO should request an ASO (and additional partner) to make the entry. If this is not acceptable to the IC, the ISO should request the IC to take back safety functions while the ISO enters the IDLH environment with a partner.

APPLYING THE ISO ACTION MODEL AT STRUCTURE FIRES

In addition to the general duties at a structure fire, the ISO Action Model can also be applied at structure fires. Remember the Four Rs: reconnaissance, resources, risk, and report? (See **Figure 13-3**.)

Risk Evaluation at Structure Fires

Although risk monitoring is a general duty, the structure fire environment calls for more specificity. For example, if an all clear is not given, firefighters (and the IC) may still assume that high risks must be taken. Other firefighters may believe that the search effort is not completed until all the living and the "dead" are found and that therefore high risk is warranted. The ISO needs to be willing to make a judgment regarding the likelihood that an un-located victim can survive the incident. To make the judgment, the ISO can evaluate fire, smoke, and building conditions to form an opinion regarding rescue profile.

The **rescue profile** is a classification given to the probability that a victim will survive the environment. Typical classifications are high, moderate, or zero for a given area in a building. In a zero rescue profile, there is obvious death or no chance for the victim to survive. Obvious zero rescue profile areas are well involved with fire. You can also read smoke to determine a zero rescue profile: Turbulent smoke, black fire, and superdense smoke issuing from a part of a building are indicators of zero rescue profile. A high rescue profile warrants additional risk taking, but remember to monitor time, smoke

rescue profile
a classification given to the probability that a victim will survive the environment; typical classifications are high, moderate, or zero for a given area in a building; in a zero rescue profile, there is obvious death or no chance for the victim to survive

THE FIRE DEPARTMENT INCIDENT SAFETY OFFICER ACTION MODEL

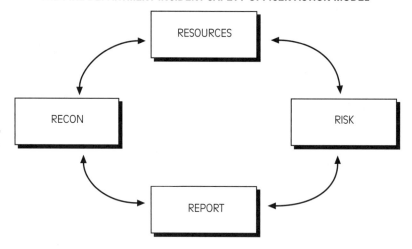

Figure 13-3 *The four Rs of the Incident Safety Officer Action Model: Reconnaissance, Resources, Risk, and Report.*

! Safety

As Chief Vincent Dunn (Deputy Chief, FDNY, retired) has said at numerous conferences: "Speed kills. Do not get caught up in the excitement of the fireground."

■ Note

Experience, knowledge, theory, study, and wisdom allow the ISO to take the visual image collected during the recon effort and make a determination, judgment, or advisement on the potential or probability for firefighter injury.

! Safety

From the ISO's perspective, the principal hazard is what, more than anything else, is likely to kill firefighters.

conditions, and firefighting effectiveness so that the risk taking can be adjusted as conditions change.

Another risk evaluation can be made regarding the "pace" of the incident. Because most of what firefighters do at a structure fire takes place in a short time window, they tend to "up the pace" for task accomplishment. A quickened pace minimizes the time to think and increases pulse rates, blood pressure, and other bodily responses. A quick pace can actually minimize the reaction time to a surprise hazard. As Chief Vincent Dunn (Deputy Chief, FDNY, retired) has said at numerous conferences: "Speed kills. Do not get caught up in the excitement of the fireground." Following an all clear, firefighters should reduce their pace. The ISO should watch for crews or individual firefighters who are out of synch with the incident pace; they are taking an inappropriate risk. At times, the ISO needs simply to state: "We are not going sixty-five miles per hour here—back it off to forty."

Recon Evaluation at Structure Fires

The recon effort at a structure fire should come early and be repeated five, ten, or even twenty times as the incident progresses (or regresses). Experience, knowledge, theory, study, and wisdom allow the ISO to take the visual image collected during the recon effort and make a determination, judgment, or advisement on the potential or probability for firefighter injury. As a starting place, the ISO uses smoke-reading and building-reading skills to define the incident environment, evaluating the environment in three dimensions: the principle hazard, environmental integrity, and the effects of the surrounding elements. In addition to evaluating the environment, the ISO should watch crew exposure to hazards. In each of these, the ISO is challenged not only to evaluate what the current condition of each is, but also to predict or anticipate the change to each of these and how each will affect the incident.

Defining the Principle Hazard   When you ask firefighters what the principal hazard of a structure fire is, they look at you and say, in a humorously sarcastic tone, "The building will burn down!" This reply is 100 percent correct. From the ISO's perspective, the principal hazard is what, more than anything else, is likely to kill firefighters. If a free-burning fire is traveling down a center hallway of an apartment building, the hazard, from the ISO perspective, is rapid fire spread that will trap firefighters (**Figure 13-4**).

The principal hazard could be the loss of integrity (like collapse potential) or a lack of a rapid egress routes from the building. Fires in void spaces, in basements, and above drop ceilings pose additional threats to working groups. Pay particular attention to these construction features—firefighter safety depends on it.

Figure 13-4 *A fire that has captured a central hallway will lead to rapid fire spread and is the principle hazard at this incident.*

environmental integrity
the state of a building, area, or condition being sound, whole, or intact

Defining Environmental Integrity   Environmental integrity is the state (sound, whole, or intact) of a building, area, or condition. The ISO is responsible for determining environmental integrity, considering factors such as weather, smoke, flame spread, and hazardous energy. At the structure fire, the ISO can define environmental integrity using the following terminology:

- Stable—not likely to change
- Stable—may change
- Unstable—may require attention
- Unstable—requires immediate attention

Classifying something as unstable is one thing, but knowing the potential *rate of change* is the important thing. Judging the rate of change (is it getting better or getting worse—and how fast?), is the ISO's foundation for effective decision making.

Defining Physical Surroundings   When getting around the incident scene, the ISO must assess the possible impact of any physical features on firefighters. The effective ISO evaluates any physical item, including terrain, foliage, curbs, posts, fences, drainages, signs, antennae, barriers, among others, and decides whether the item could affect the operation.

One physical feature—a sloping grade—has been cited as a significant factor in multiple firefighter deaths. In sloping grade incidents, a crew that thinks they are working on the first floor may in fact be working on the second

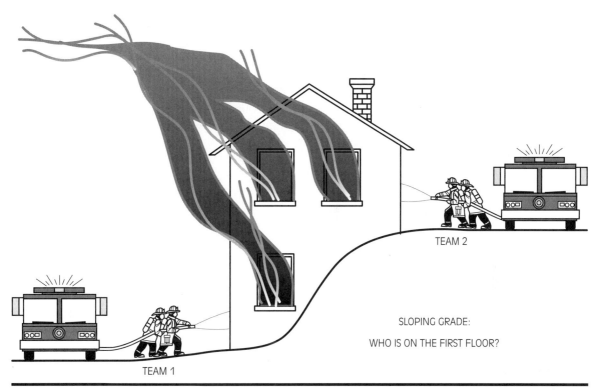

Figure 13-5 *Sloping grades may cause dangerous miscommunication. In this figure, both team 1 and team 2 may believe they are on the first floor.*

■ **Note**
The effective ISO evaluates any physical item, including terrain, foliage, curbs, posts, fences, drainages, signs, antennae, barriers, among others, and decides whether the item could affect the operation.

(**Figure 13-5**). Usually, the presence of a fence or other barrier masks this situation from the IC or the advancing crews. When observing sloping grade situations, the ISO should relay the information to the IC and suggest that it is relayed to all responders. The IC should relay how the varying levels are to be labeled and communicated to improve clarity for those who need to report their location.

Fences, shrubbery, parked vehicles, security hardware, and other barriers may impede incident operations by restricting access or impairing an adequate size-up. The ISO should determine the relevance of the barriers to the incident as a whole. As a rule, the ISO should survey the entire area of incident impact and ensure that each crew has at least two escape areas (safe havens), as well as a clear corridor for egress.

Crew Exposure to Hazards   While evaluating the environment, the ISO should also look at the crews. The relationship can be expressed as:

Principal hazard ± Integrity + Physical hazards + Crew activity
= Crew exposure

While the equation seems simple, it is often overlooked because the crew is focused on a task and may not see the big picture (they are in a linear thinking mode). The ISO looks to see if the equation adds up to a negative outcome. Often, *what* the crews are doing and *how* they are doing it make the equation negative. Here are some considerations:

- *Tool versus task:* Are tools being used correctly? Are they the right ones for the task? Are they being stressed past their intended design limits? Firefighters are amazingly creative and results oriented. Their attributes are admirable but can lead to injury.

- *Team versus task:* Are there too few firefighters for the task? Are there too many? If too many people are involved in a task (a "magnet" task— one that everyone wants to be involved in), injury may result due to congestion and team immobility **(Figure 13-6)**.

- *Rapid withdrawal options:* Does the crew have an escape path if things go wrong? Are there multiple escape options? How far does the team have to travel to escape?

- *Rapid intervention options:* Is a rapid intervention crew or company (RIC) available? Does the RIC have a solid understanding of scene hazards, ingress options, and hazards? RIC is required for all IDLH structure fires. The ISO should continually relay access, egress, and hazard information to the RIC. Most of us realize that rapid intervention is *rarely rapid.* Account for this reality.

Figure 13-6 *Having too many people perform a task is just as dangerous as having too few.*

Figure 13-7   *Can you spot the trip or fall hazards in this typical overhaul operation?*

- *Trip/fall/struck-by hazards:* Evaluate the trip and fall hazards you find. Can something suddenly let go and strike a crew? Look at **Figure 13-7** and count how many trip and fall hazards you spot. Compare your evaluation with **Figure 13-8**. This is only a simple overhaul operation, but a simple "Be careful, look at the trip and fall hazards here" can make the team leader aware and perhaps more focused.

Resource Evaluation at Structure Fires

The ability to evaluate, manage, and assign resources is often used to measure the effectiveness of the incident commander. Budgetary concerns, response times, politics, and compatibility issues combine to limit the availability of resources to overcome an incident. While having unlimited resources to handle a serious incident is ideal, it is also unrealistic to expect of most communities. The ISO must look at what *is* available and determine whether it is adequate. Generally, time, personnel, and equipment are the key resource

FALL AND TRIP HAZARDS

1. UNSUPPORTED LADDER
2. UNATTENDED HOSE AND NOZZLE
3. DOORWAY WITH NO LADDER OR STAIR
4. CONSTRUCTION DEBRIS
5. HOSELINE UNDERFOOT

6. WORKING IN DITCH OR ACCESS CUT
7. WILL WALL SUPPORT LADDER? (ROOF IS GONE)
8. FIRE AX
9. CONSTRUCTION GRADE (NOT SMOOTH, MANY RUTS)

Figure 13-8   *Did you spot more than nine?*

considerations. With each of them, the ISO's basic question is, "Do we have enough to do what we're trying to do?" Answering this question acts as a check and balance that can help the IC.

Time   Time is often a forgotten resource. From the ISO's perspective, time is an important resource that dictates hazard priorities, permits reconnaissance efforts, or makes the difference between an imminent threat to life or a minor operational concern. ISOs must have a keen sense of time as a resource: Time cannot speed up and it cannot slow down. You cannot trade it or buy it. You

Figure 13-9 *Dispatch can help time management by announcing time in 5- to 15-minute intervals.*

cannot keep it or give it away. Time does, however, tend to "slip away" from time to time. (Pun intended!)

Unfortunately, our *perception* of time changes at working incidents. Most fire officers can recall an event where they would swear they were working for only 10 minutes when in fact it was 30. The inverse has probably been experienced also. How can the ISO evaluate and use time effectively? One suggestion is to preplan the use of a dispatch or communication center (**Figure 13-9**) to periodically broadcast the transpired incident duration over the assigned radio frequency. If such a procedure does not exist, the ISO should use a stopwatch, an egg timer, or other device to monitor the passage of time.

The ISO should also project on-scene time. Obviously, a 2-hour firefighting effort has different needs then a 12-hour conflagration. Short-duration incidents (less than 2 hours) may require just rest, active cooling, health monitoring, and hydration. Longer incidents may indicate the need for nourishment, extended rest, fresh crews, and sanitation logistics. Projecting on-scene time can also remind the ISO that either darkness or the usual afternoon thunderstorm is coming.

Reflex time is also a consideration. One truism seems to fit the fire service: If you need it, and it is not there, it is too late. The ISO should look to see if staged crews are available and ready for an assignment and project their reflex time if a surprise were to upset the action plan. Even though the ordering and assignment of resources is an IC function, the ISO can initiate a subtle reminder if necessary.

Personnel   The ISO should evaluate whether enough personnel are available for the incident action plan and make sure that enough responders are in staging, should an immediate need evolve. If not, the ISO should bring the observation and solution as a suggestion to the IC, along with an evaluation of the number of responders exposed to hazards. Maybe too many people are taking too much risk.

Equipment   An understanding of the types of equipment present, what is needed, and what is on the way can help the ISO evaluate the adequacy of equipment for the fire. In most cases, the issue is one of "fit." Does the equipment being utilized *fit* the action plan? Does it all reach?

Reach applies to more than hose and ladder length. Reach can be applied not only to ladders and hose lines, but also to radio communications, fire streams, and building depth. We typically look at reach as a height issue, but it also applies to the distance firefighters must travel *into* a building. Operations taking place well within a building (over 300 feet) present unique concerns regarding air use, hose line effectiveness, equipment shuttling, lag time, and rapid egress.

Report Issues at Structure Fires

At structure fires, ISOs should follow the 15-minute rule for face-to-face communications with the IC, during which risk, recon, and resources are the topics of discussion. ISOs can also use these opportunities to update checklists, forms, diagrams, and other documents. They should consider developing a safety briefing sheet that can be passed to the staging manager for multiagency responses, longer-duration incidents, and unusual fires. Safety briefings seem to be underutilized at structure fires—what a disservice to responders! Safety briefings have many benefits: They show responders that their safety is a priority and help them get dialed into the incident prior to assignment.

UNIQUE CONSIDERATIONS AT STRUCTURE FIRES

Residential Versus Commercial Fires

Unfortunately, classifying building fires as residential or commercial can set up firefighters with mistaken expectations. In reality, some residential properties can exceed the size and fire load of a commercial building, and some commercial buildings present minimal hazards to firefighters. Size is not the only issue. Building use is equally important. Many buildings were built for residential use but are used as commercial structures—and the converse is also true. ISO should therefore weigh building size and use when making a "read" on the situation. Are residential tactics being used on a residential property that actually requires more commercial-appropriate tactics (more attention to ventilation and greater GPM)? Although these are IC and action plan issues, the ISO should be alert to the traps of classifying a building as residential or commercial.

Buildings with Central Hallways and Stairwells

A center-hall building or one with a central stairwell creates unique problems for firefighters. From the ISO perspective, these buildings present a tremendous risk for the advancing firefighter: The path for ventilation and fire spread is the very same path that firefighters need to perform search and fire attack. Dense smoke in hallways and stairwells become another risk: Smoke is fuel that ignites when the right mixture and temperature combine, causing a rapid fire spread that is faster than firefighters can react. The key is ventilation; the tactics employed *must control* smoke and heat. Ventilation in center-hall/ stair buildings is the number one tactical priority (for search, fire attack, and firefighter safety). These are IC tactical issues, but, if they are not adequate (read the smoke and read the building), the ISO needs to intervene.

> **Safety**
> Ventilation in center-hall/stair buildings is the number one tactical priority.

Strip Malls

The neighborhood strip mall can be a firefighter killer. Specifically, high fire loads, common ceiling spaces, long open-span trusses, and decorative façades can turn the simple DVD rental store fire into a surprisingly nasty firefight. Watching smoke can pay huge dividends for the ISO; be the eyes for the incident commander on strip mall fires. As firefighters attempt to confine the fire, watch for brown smoke, under pressure, coming from the façade distal to the fire attack. Fire can spread rapidly in façades, accompanied by rapid collapse. Often, crews use the space below a shared façade for attack access; pay particular attention to these (**Figure 13-10**). Make sure the occupancies on either side of the involved unit have their concealed spaces exposed and preventilated. If these tactical priorities are not in place, share a solution with the IC. Strip malls typically have heavy roof loads, like HVAC systems, grease hoods, signs, satellite dishes, and other equipment. Stand-alone fast-food restaurants can share the same hazards as a strip mall.

Remember also that strip malls and stand-alone fast-food restaurants are relatively inexpensive to build. A fire that warps the bar trusses in a single unit may lead to a decision to completely demolish the building. The developer can "scrap" the entire strip mall for less than it costs to fix a single unit damaged by the fire. Practice sound risk management at strip mall fires.

> **Note**
> The developer can "scrap" the entire strip mall for less than it costs to fix a single unit damaged by the fire. Practice sound risk management at strip mall fires.

High-Rise Buildings

Fighting high-rise fires is a science unto itself, and the ISO-related concerns could take up another whole book. The intent here is to give some general direction for the ISO at these complex incidents.[1] First, and perhaps

Figure 13-10 *Note the fire spread and collapse in this strip mall facade. Firefighters made a good stop.*

foremost, the ISO should request one or more ASOs for growth-stage fires in high-rises.

Let's look at some other considerations:

ISO Functions at High-Rise Incidents The ISO should take up a position with the command post. If the IC is in the lobby, be there. If command is two floors below the fire, be there. If command is on the street, be there. The reason is simple: ISOs need to be in on action planning and able to hear what building reps and technical specialists have to say. They should be ready, however, to bounce between command, liaison, and plans. In all cases, they need to remain "strategic" in their safety thinking. Of particular interest to ISOs should be:

- The overall action plan
- Control of building systems, such as elevators, HVAC, and stairways
- Occupant evacuation or shelter-in-place issues, a safety issue with firefighters going one way and occupants the other
- Communication with and feedback from the assigned ASOs

ASO Functions When command takes a position in the lobby or street, an operations section is typically set up on the upper floors below the fire. Assign an ASO to go work with Ops, possibly bouncing between the ops section chief location and the floors below, where rehab and forward staging are located.

You may also need an ASO to perform recon outside the building, monitoring access, traffic, hose, and glass/debris issues. The ASOs need to deal with typical safety functions associated with recon, tactical-level risk, and rehab. Of particular interest to the ASO should be:

- *The physical demands on firefighters:* Shuttling equipment in itself taxes the firefighters. Firefighters on the fire floor experience more rapid signs of heat stress. Rehab is hugely important: Active cooling, hydration, and electrolyte and nourishment replacement are vital right away!

- *Internal traffic control:* A plan should be in place and communicated for firefighter travel routes and accountability check-in.

- *Compartment integrity:* The ASO should encourage firefighters to be the eyes and ears for compartmental collapse possibilities. Dropped ceiling grids collapse early. More importantly, crews need to report the condition of stairwell cracks, stuck or bound exit corridor doorways, and window glass.

- *Establishing no-entry zones around lost windows:* Air (and smoke) pressure variances between the building and the outside atmosphere may create a jet vortex near lost windows that can pick up and move an unsuspecting firefighter.

- *Development and delivery of safety briefings:* This may be an inside and/or outside ASO task.

- *Outside issues:* The ASO outside the building needs to be particularly observant of traffic issues; exposure to dropping glass, debris, or humans; and maintenance of established zones.

SUMMARY

At structure fires, the ISO must apply a systematic approach that takes into account general ISO functions, action model components, and unique considerations. Specific to the general duties, the ISO should monitor risk and operational effectiveness to make sure they fit with incident benchmarks and the established action plan. The ISO can apply "reading skills" to determine fit. When accountability issues arise, the ISO should consider asking the IC for a PAR, and effective PAR communicates not only the number of responders, but also the assignment, location, and conditions. Labeling zones as IDLH, no-entry, collapse, and support can minimize interpretation issues. Other general duties include the evaluation of traffic—especially

apparatus movement—and the recognition and need for ASOs.

The ISO can adopt the ISO Action Model to structure fires. For the risk component, the ISO should evaluate rescue profile and crew pace, the two factors that can throw an incident into a negative risk situation. For recon, the ISO needs to evaluate and define three environmental dimensions: the principal hazard, integrity, and the physical surroundings. Fires in buildings that are located on a sloping grade present miscommunication issues that can be disastrous. The ISO can use the three environmental dimensions to judge crew exposure. Teams, tools, tasks, withdrawal options, and trip/fall hazards can all lead to negative outcomes for crews.

Time, personnel, and equipment are the re-source elements that need to be judged by the ISO. Adequacy is the primary issue: Does it all *fit*? The ISO cannot lose track of time and should project the on-scene time to help fore-cast hazards and needs. Safety briefings (part of the report component of the action model) are underutilized at structure fires.

The classification of fires as residential ver-sus commercial can lead to inappropriate tac-tics. Building size and use are better ways to classify incidents. Center-hallway/stairwell, strip mall, and high-rise buildings hold unique hazards that require ISO attention. At high-rise fires, the ISO should request assistance so that specific safety functions can be addressed.

KEY TERMS

Cold zone   Establishes the public exclusion or clean zone. There are minimal risks for human injury and/or exposure in this zone. Denote cold zones with green tape.

Collapse zone   Areas that are exposed to trauma, debris, and/or thrust of a collapse; a more specific form of a no-entry zone.

Environmental integrity   The state of a build-ing, area, or condition being sound, whole, or intact.

Hot zone   The area presenting the greatest risk to members and often classified as an IDLH atmosphere. Denote hot zones with red tape.

IDLH zone   Areas in or around the building where working firefighters are exposed or may become exposed to smoke and heat. Persons working in an IDLH zone shall have a partner, work under the two-in/two-out rule, and be tracked through an accountability system.

No-entry zone   Areas where no person— including firefighters, police, other respon-ders, or the general public—should enter due to the serious or unpredictable nature of

a hazard or condition. Denote no-entry zones with red/white chevron tape.

PAR (personnel accountability report)   An organized reporting activity designed to ac-count for all personnel working an incident. To be truly effective, PAR radio transmis-sions should include assignment, location, and number of people in the assignment.

Rescue profile   A classification given to the probability that a victim will survive the environment. Typical classifications are high, moderate or zero for a given area in a build-ing. In a zero rescue profile, there is obvious death or no chance for the victim to survive.

Support zone   Areas where firefighters, other responders, IMS staff, and apparatus are op-erating or staged. The general public should *not* be allowed to wander into the support zone.

Warm zone   A limited access area for mem-bers directly aiding or supporting operations in the hot zone. Denote warm zones with yel-low tape.

POINTS TO PONDER

Roof Collapse

On September 14, 2002, a 53-year-old male career firefighter died after falling through a roof following roof ventilation operations at a house fire.

(continued)

(*continued*)

The structure was a 96-year-old, 2½-story wood frame dwelling with balloon frame construction. The roof was steeply pitched and had intersecting gables consisting of 2 × 4-inch timbers covered with 1-inch wood planks and four or five layers of asphalt shingles. The building was located on a corner lot on a grade. The second floor was accessed at street level from the rear. The first-arriving engine reported smoke showing under the rafters and established command per department policy. The IC and engine crew conducted an interior size-up to determine the location of the fire and whether there were any occupants (there were none). Because of the heat and smoke encountered as they tried to climb to the top floor, the crew had to back down the stairs. The IC then exited the structure and called for roof ventilation. A firefighter from a rescue squad replaced the IC on the second interior attack attempt. A second interior fire attack was begun through the rear (second floor) of the building, which initially was relatively free of smoke. As this attack crew tried to climb the stairs to the top floor, they too encountered heavy smoke and extreme heat, causing them to back down the stairs to wait for ventilation.

After positioning the aerial platform over the roof reportedly "as far as it would go," the victim and his partner, carrying only a chain saw, exited the aerial platform and walked approximately 15 feet to the area to be ventilated. The victim's partner had donned his self-contained breathing apparatus (SCBA) before leaving the ground and, upon exiting the platform, went on air due to heavy smoke. The victim, who was not wearing SCBA, was observing his partner making the ventilation cuts. After the last cut was made, the victim, who had been covering his face with his hands (presumably because of the thick smoke), told his partner that they had to leave immediately. The firefighters retreated toward the aerial platform, but the victim stopped a few feet from the platform, saying he could not continue. Seconds later, the area of the roof under the victim failed, and he fell through the roof into the structure and the fire. Within minutes the interior attack crew found the victim and, with the help of the rapid intervention crew (RIC), removed him. He was transported to a local hospital where he was pronounced dead.

The victim was on the roof for less than 7 minutes. He did not fall through the hole cut by the saw but rather through an area of the roof between the cut hole and the aerial platform. According to the fire department report, this area presumably failed due to direct exposure to the fire below and the weight of the victim above. The victim fell approximately 10 feet.

For Discussion:

1. From the information given, what are some of the structural (building) factors that contributed to the collapse?

2. What information presented in the case could have been used to establish a rescue profile for potential victims on the various floors and within various rooms of the building?

3. What was the role of *elapsed time* at this incident?

4. From an ISO perspective, what descriptions from the report indicate an imminent threat requiring action? Which descriptors indicate a potential concern?

Note:

This case study was taken from NIOSH Firefighter Fatality report #2002-40 available at www.cdc.gov/niosh/fire.

REVIEW QUESTIONS

1. Discuss the relevance of risk taking to incident benchmarks.

2. What two factors can help in judging operational effectiveness at structure fires.

3. An effective PAR should include the communication of what four elements?

4. What is the essential difference between an IDLH and no-entry zone?

5. List four times when an ISO should request ASO assistance at structure fires.

6. What is meant by a zero rescue profile?

7. List the three dimensions that need to be defined during environmental reconnaissance.

8. What is a magnet task?

9. Why should on-scene time be projected by the ISO?

10. List the three resource considerations at structure fires.

11. What is the trap in labeling structure fires as residential or commercial?

12. List several unique hazards at strip mall structure fires.

13. List four ISO functions unique to high-rise fires.

14. List six ASO functions unique to high-rise fires.

ADDITIONAL RESOURCES

Clark, William E. *Firefighting Principles and Practices,* 2nd ed. Tulsa, OK: Fire Engineering Books and Videos, a division of PennWell, 1991.

Emery, Mark. "13 Incident Indiscretions." *Health & Safety for the Emergency Service Personnel* 15, no. 10 (October 2004).

Emery, Mark. "The Ten Commandments of Intelligent and Safe Fireground Operations." *Health & Safety for the Emergency Service Personnel* 15 no. 11 (November 2004).

Fire Fighter Close Calls. Web site containing fireground stories, near miss reports, and lessons learned at structure fires: Available at: www.firefighterclosecalls.com.

Mittendorf, John W. *Truck Company Operations.* Tulsa, OK: Fire Engineering Books and Videos, a division of PennWell, 1998.

Norman, John. *Fire Officer's Handbook or Tactics,* 3rd ed. Tulsa, OK: Fire Engineering Books and Videos, a division of PennWell, 2005.

NOTES

1. Author's note: A special thanks to Captain Fred McKay and Safety Chief David Ross (Toronto Fire Services), as well as to Battalion Chief Gerald Tracy (FDNY), for sharing their wisdom on high-rise fires. The author takes all responsibility for interpreting their input.

Chapter

14

THE ISO AT WILDLAND AND INTERFACE FIRES

Learning Objectives

Upon completion of this chapter, you should be able to:

- List and describe four incident types that can be applied to the wildland fire.
- List the three factors that influence fire spread, and define blow-up and flaring.
- List the leading stresses requiring rehab at the wildland fire, and describe the types of behaviors that would indicate effective rehab efforts.
- List four situations that may require the appointment of an ASO at the wildland fire.
- List the three most common principal hazards at a wildland fire.
- Define LCES.
- Discuss a troubling issue that may arise when ground firefighters interface with aircraft.

INTRODUCTION

The phrase "wildland and interface fires" in the chapter title is not meant to be restrictive. A farmer's wheat stubble fire is not necessarily a wildland fire by definition; it is a cultivated field fire. Likewise, not all wildland fires are interface fires. The term "interface" is usually applied to the use of structurally oriented firefighters and equipment to prevent the spread of vegetation fires to structures. Given that, in this chapter, for brevity's sake, we use the generic terms "wildland fires" or "wildfires" to apply to all such incidents.

At times, structural firefighters treat wildland fires as a sort of "nuisance" event that lacks the challenge and excitement of structural fires. Too often, the attitude that accompanies the nuisance incident can result in reduced attention to personal protective equipment, adequate fire flow, and sound attack strategies. Conversely, trained wildland firefighters believe that wildfires have more challenges and excitement than structural fires and thus put more emphasis on PPE and sound attack strategies. When the two groups came together for a common fire, the differences present interesting conflicts. Thankfully, most of us learned the lessons of the California firestorms, the Long Island (New York) wildfires, and the vast wildfires in Florida, Oklahoma, and Texas. More than ever, wildland and structural firefighters are sharing important perspectives and finding commonality in operations.

The scope of this chapter is to provide the structurally oriented fire officer with ISO insight as part of the initial response to a wildland fire. As a wildland fire grows, so does its resource demand. At some stage, the fire may grow beyond the resources of the local fire department and its regional mutual aid assistance. Once this occurs, the ISO functions are likely to be transferred to a trained safety officer who is part of an incident management team (IMT). In this chapter, we therefore address the *initial ISO efforts,* and, because the language may be different, we try to bring clarity by addressing the ISO duties at *Type-5* and *Type-4* wildland fire incidents (see The Wildland Fire Typing Scheme).

The Wildland Fire Typing Scheme

The wildland fire community has used typing schemes for decades. Typing is a way of creating standardized language and understanding of resource capabilities (covered in Chapter 11).

The generally accepted definitions of incident types are based on the level of expertise and equipment required. An example of the size of a wildland fire is included for each of the following incident types.

Type-5: Local, agency, or jurisdiction specific

Small wildland and interface fires that can be contained and extinguished with local initial responders

(continued)

(continued)
Type-4: Multiagency or jurisdiction

Wildland and interface fires that exceed the resource capabilities of initial responders but are manageable with additional resources

Type 3: Regional

Wildland and interface fires that exceed local resources and require regional mutual-aid resources and specialized state and federal resources for containment and extinguishment

Type 2: State

Large wildland and interface fires that require the activation of statewide fire plans and the coordination of a significant number of specialized resources that may include those from the federal level

Type 1: National

Large campaign-type fires that require the coordination of a significant number of interstate and federal resources

ISO GENERAL DUTIES AT WILDLAND FIRES

Safety

Upon arrival and assignment at a wildland fire, the ISO should quickly grasp the potential for firefighters being overrun by the fire.

Upon arrival and assignment at a wildland fire, the ISO should quickly grasp the potential for firefighters being overrun by the fire. The well-known factors that effect fire spread are weather, topography, and fuels. Some general considerations for each are as follows:

- *Weather:* Temperature, relative humidity, barometric pressure, winds, and weather "events" such as microbursts and tornadic activity need to be considered.
- *Topography:* Factors such as slope (degree), aspect (relationship to the sun), and physical features (chimneys, saddles, barriers, etc.) all influence wildland fire behavior.
- *Fuels:* Additionally, wildland fuels are affected by moisture content, fuel type (ground, aerial, etc.), and continuity.

These factors may combine in such away that firefighters could be overrun, as in a rapid fire spread in a structure fire. Also, just like a flashover in a structure, wildland fires experience hostile events:

- *Blow-up:* A wildland fire term used to describe the sudden advancement and increase in fire intensity due to wind, prewarmed fuels, or a topographic feature such as a narrow canyon or "chimney." Sometimes the word "**blow-up**" is used when a ground or surface fire becomes an aerial or crown fire. Blow-up is so named because it often disrupts or

blow-up
a wildland fire term used to describe the sudden advancement and increase in fire intensity due to wind, prewarmed fuels, or topographic features, such as a narrow canyon or a chimney

changes control efforts. Because of the sudden increase, a violent convection column causes additional concerns and usually accelerates fire intensity.

flaring
a sudden, short-lived rise in fire intensity, attributed to wind, fuel, or topographical changes. Flaring can be a warning sign of an upcoming blow-up

- *Flaring:* A sudden rise in fire intensity that is short-lived. **Flaring** can be attributed to wind, fuel, or topographical changes. Flaring can be a warning sign of an upcoming blow-up.

The ISO needs to evaluate the many factors related to fire behavior and then attempt to classify the fire in terms of firefighter safety. Evaluating flame length can also help the ISO determine if initial attack crews are at risk (see Wildfire Flame Length Interpretations).

Wildfire Flame Length Interpretations

Less than 4 Feet: The fire can generally be attacked directly using hand lines and tools.

4 to 8 Feet: The fire is too intense for a direct attack on the head. A flanking attack with increased GPM may be effective. Indirect firebreaks and wet lines are advisable.

8 to 11 Feet: The fire presents serious control problems. Direct fire attacks are dangerous.

Over 11 Feet: Major fire runs are likely. Defensive measures are required.

By evaluating wildland fire behavior, the ISO can determine the potential of the incident. Initial observations and concerns should be shared with the IC.

Monitoring Issues at Wildland Fires

Risk The nature of wildland fires tends to cause victims to self-rescue. That is not to say there will not be victims and this possibility needs to be determined. Once the victim question is resolved, the prevailing life risk is to firefighters. The incident action plan should reflect this change in a reduced-risk profile for the action plan.

Operational Effectiveness At most wildland fires, the ISO is not in a good position to evaluate operational effectiveness because of distance, terrain, and smoke. The use of roving ASOs may be helpful where the geographical layout of the incident minimizes ISO effectiveness. Even still, the ISO needs to make a judgment that the actions underway are making progress toward achieving the objectives of the IAP. Taking a tour in a vehicle or helicopter can help.

Personal Safety System Issues at Wildland Fires

Accountability Systems   The collection of accountability passports may prove to be troublesome in the initial stages of the wildfire. There have been too many cases of firefighters unaccounted for—until their deaths are discovered. During attack operations, firefighters may spread out and lose the typical crew integrity found at structure fires. Sometimes, the ISO may need to intervene and close up the ranks.

Control Zones   It is rarely practical to add typical control zone language (hot, warm, and so on) for the wildland fire. Many of the zones used for wildfires are based on the descriptive parts of the wildfire. Words such as the "head," "flanks," "origin," "spots," and the like are used to describe parts of the wildfire. Occasionally, geographical feature names, like "saddle," "chimney," and "interface," are used to describe zones. Examples of wildland parts and features can be found in **Figure 14-1.** Firefighters working between the head/flanks

Figure 14-1   *Various parts of the fire or geographical descriptors are used to label zones at wildfires.*

● **Caution**
Firefighters using structural PPE for wildland incidents are at extreme risk for heat stress.

! **Safety**
Firefighters should not be allowed to go back to work based on their perceived comfort; a qualified medical attendant should make the determination based on core temperature and other vital signs. Physically stressed firefighters *should not* be allowed to leave the incident; they should remain in rehab until they have been medically cleared or are transported to a definitive care facility.

and spots could be endangered. Threatened structures can be classified in numerous ways but generally get classified as defensible or indefensible. The portion of the wildland where fire has already past is called the *burn* or the *black*. Generally, the black is a safe zone for firefighters, although CO exposure and reburn are risks (**Figure 14-2**). Anecdotally, a significant reburn took place at the 2000 Bobcat Fire near Loveland, Colorado. The fire burned through an area on one day then burned back through the area, more intensely, the next day. Establishing resource staging areas, the command post, and support areas in areas upwind and protected by firebreaks should be considered when evaluating control zones.

Rehab   Exposure to heat, smoke, and physical exertion lead the list of factors that require significant attention to rehab. Firefighters using structural PPE for wildland incidents are at extreme risk for heat stress. Many structural fire departments now supply wildland-appropriate gear for their firefighters (if not, the department is willingly setting firefighters up for a heat stress injury when structural PPE is used). In all cases, rapid hydration and electrolyte replacement are essential; make sure rehab has this covered. Medical monitoring should also be prioritized for rehab; heart attacks remain the number one killer of firefighters. Firefighters should not be allowed to go back to work based on their perceived comfort; a qualified medical attendant should make the determination based on core temperature and other vital signs. Physically stressed firefighters *should not* be allowed to leave the incident; they should remain in rehab until they have been medically cleared or are transported to a definitive care facility. Rehab attendants should consider cardiac monitoring for any firefighter who exhibits even remote signs of heat stress. If advanced life support medical monitoring is not available at rehab, consider asking the IC to order it to monitor those with signs of heat stress.

Figure 14-2
Generally speaking, the black is considered a safe zone, although CO exposure and reburn are risks.

Defining Other Needs at Wildland Fires

Traffic   Smoke obscuration is the leading traffic concern at wildland incidents. Efforts to divert traffic away from smoky areas are warranted. Small fire apparatus dispersed over a wide geographical area can create daunting concerns. Drivers of small, mobile brush patrol vehicles may not be able to see or hear firefighters working on foot in the area. A general safety message broadcast over the radio frequency can remind drivers to use spotters (especially when backing) and to back off speed. As with structure fires, remind firefighters to "look both ways" and never to run or jog around in apparatus movement areas.

> ! **Safety**
> A general safety message broadcast over the radio frequency can remind drivers to use spotters (especially when backing) and to back off speed.

Need for ISO Assistance   The ISO should request an assistant safety officer(s) at wildland fires in certain circumstances:

- Fires that impact a widespread geographical area
- When a plans section is established (The ISO needs to provide input on action plan changes and strategic-level issues, and ASOs are needed to accomplish the ISO field component [recon, rehab evaluations, etc.].)
- Fires that are active for more than four hours
- Anytime a base camp is established (The ISO needs someone to communicate the action plan and safety briefings to incoming crews. The base camp may also be the rehab area, where food service and sanitation needs are established, services that need to be evaluated.)

APPLYING THE ISO ACTION MODEL AT WILDLAND FIRES

Risk Evaluation at Wildland Fires

Evaluating risk in the wildland fire environment requires special considerations. It is easy to say that no risk should be taken if evacuation and search efforts have eliminated rescue needs, but do you write off a farmer's field or let a watershed area burn? Value has to be considered. Good fire control efforts can actually prevent millions of dollars in mud slide damage when the rains come after a fire or save hundreds of thousands of dollars in agricultural revenue. The effective ISO understands these concerns and works with the IC to achieve a "calculated" risk-taking environment that favors firefighter safety and is still aggressive in fire control efforts. In other words: *intellectual aggressiveness.*

Judging the pace of the incident is part of risk monitoring. The time window for initial effectiveness at a wildland fire is usually measured in minutes and hours, as opposed to seconds and minutes, as at the structural fire. Structural firefighters engaged in wildland fires should back off from their typical pace. This statement is certainly arguable if structural protection, with its typical short time window, is necessary in interface fires. Another viewpoint offers a variant: Making a structure defensible against an advancing fire could take 20 to 30 minutes. If that time is not available, given flame lengths and speed of fire spread, then the structure is *not* defensible.

If the fire is progressing faster than crews are effective, the IC may become trapped in linear thinking and regress to habitual behaviors in crisis mode. The ISO who senses this development must shore up the IC function with cyclical thinking, suggestions, and support.

■ **Note**
Making a structure defensible against an advancing fire could take 20 to 30 minutes. If that time is not available, given flame lengths and speed of fire spread, then the structure is *not* defensible.

Recon Evaluation at Wildland Fires

The recon effort at a wildland fire is not as simple as walking around the building. Recon vehicles, helicopters, and ASO field reports all help. ASOs or lookouts (trained in ISO functions) can help with recon, and their reports are important not only to the ISO but to the IC as well. Coordination is key. Caution: Climbing to high ground to get a good look at the fire can prove fatal. If an elevated feature affords a good sight line, take advantage of it as long as there is no remote possibility that the fire will go there. Keep in mind that the purposes of recon are to define the principal hazard, judge the potential for environmental change (integrity), define the impact of the physical surrounding, and equate the exposure of crews to the environment and principal hazard.

● **Caution**
Climbing to high ground to get a good look at the fire can prove fatal.

Defining the Principal Hazard   The principal hazard in most wildland environments can result from one of three causes: rapid fire spread, traffic issues, or physical exertion. These assumptions are well based in statistics. Intervention is usually required in each area to prevent injuries.

■ **Note**
The principal hazard in most wildland environments can result from one of three causes: rapid fire spread, traffic issues, or physical exertion.

Defining Environmental Integrity   Environmental integrity considerations include factors such as weather, smoke, flame spread, and hazardous energy. Just like the structure fire, the ISO can define environmental integrity using the following system:

- Stable—not likely to change
- Stable—may change
- Unstable—may require attention
- Unstable—requires immediate attention

When judging the potential for fire spread overruning firefighters, do not downplay the potential of hazardous energy. Downed wires and ground transformers can create a ground gradient or direct electrocution that can zap unsuspecting firefighters. Include this reminder in safety briefings.

Defining Physical Surroundings   The wildland fire environment is as diverse as nature itself. Numerous trip and fall hazards top the list, and animals and vermin may present hazards not routinely experienced by structural firefighters. Does everyone have good footwear?

Burned-out trees and dead limbs, known as snags, may have been weakened and are a real hazard to firefighters. *Look up and live* applies to power lines *and* snags. Reminders are usually welcome.

Crew Exposure to Hazards   The equation for crew exposure remains the same:

Principal hazard ± Integrity + Physical hazards + Crew activity
= Crew exposure

Linear thinking sometimes blinds firefighters to environmental hazards. Remember, it is *what* the crew is doing and *how* they are doing it that makes the equation negative. Some considerations follow:

- *Tools:* Chain saws are invaluable, and they create dangers to those not accustomed to their operation for clearing trees, brush, and other obstacles. A structural firefighter may be quite accomplished at using a chain saw for cutting roofs, but a 4-foot diameter tree or snag requires different skills.

- *Team versus task*: Crews protecting structures may get sucked into structural firefighting if the wildland fire penetrates a building. If so, inquire as to whether the crews have switched to structural PPE and are following the two-in/two-out rule.

- *Rapid withdrawal options:* The acronym "LCES" is used to address rapid withdrawal options. **LCES** stands for *l*ookouts, *c*ommunication methods, *e*scape routes, and *s*afety zones. Individual crews should be constantly pointing out LCES options. Some LCES advocates have altered the acronym by adding *Awareness*—making the acronym *LACES* (pronounced as in boot laces). If you do not hear LACES chat on a regular basis, the crew leaders are not doing their job, and a friendly (and private) reminder is warranted. Once the friendly reminder is delivered, gauge the reaction of the leader. A negative reaction may indicate that the leader needs chemical balancing and that rehab efforts are behind the curve.

Backing apparatus into dead-ends, driveways, and structural protection assignments is an important rapid withdrawal option. This

LCES
an acronym that stands for *l*ookouts, *c*ommunication methods, *e*scape routes, and *s*afety zones

Safety
Backing apparatus into dead-ends, driveways, and structural protection assignments is an important rapid withdrawal option. This is not common practice for municipal-oriented firefighters, but it is essential for rapid retreat in the wildfire environment.

is not common practice for municipal-oriented firefighters, but it is essential for rapid retreat in the wildfire environment.

Resource Evaluation at Wildland Fires

Time   Projecting on-scene time can remind the ISO of many considerations, especially the advent of nightfall and predicted weather changes. Make sure you have a good weather forecast and use your weather-watching skills (Chapter 9). Travel distances can affect reflex time for immediate resource needs; have multiple staging points for equipment at a geographically widespread incident. Fit firefighters can probably work hard for 10- to 12-hour work periods (with effective rehab). Less fit firefighters may not be able to last that long. Active firefighting operations are typically suspended or become conservatively defensive at dark because there are just too many unseen hazards at night.

Personnel   The ISO should evaluate whether enough personnel are available for the incident action plan and whether enough responders are in staging should an immediate need evolve. If not, the ISO should bring the observation and suggested solution to the IC.

Spirits should be high. Structural firefighters working a challenging, tough, gritty, and taxing interface fire should be *happy!* You should feel excitement and contagious enthusiasm among the crews. You should observe examples of humor, trivial (fire house type) complaining, and motivational desire. Crews should be jockeying in staging for peach assignments. All of these indicators should be present with an underlying sense of focus (**Figure 14-3**).

Some may argue this point, but it is true. Most firefighters live for fires, structural or otherwise. They know that wildland firefighting is physically tough and requires concentration and awareness. The adrenaline should be throbbing! Once the initial "rush" has worn off, firefighters should glide into a routine. If rehab is effective, the routine *should still be energetic* (with all the positive indicators). If not, rehab is likely failing and needs adjustments (see Chapter 10). Making this observation is an incredibly important duty of the ISO when evaluating personnel at the wildland fire.

Equipment   Structural firefighters may not be accustomed to lengthy hose lays, water supply issues, and the use of Class-A foam in wildland environments. Pump calculations may be off—putting firefighters at risk. Some apparatus may not have hose bed protection; flying embers can ignite stored hose. Again, include reminders in safety briefings and interface with crews during recon to create awareness of these safety issues.

Figure 14-3 *The ISO should observe a sense of focus among crews.*

Report Issues at Wildland Fires

For initial operations at wildland fires, the 15-minute rule for face-to-face communication with the IC is appropriate. In prolonged incidents, the schedule can be amended. The evaluative results of risk, recon, and resources are the topic of discussion for the face-to-face updates. ISOs can also use them to update checklists, forms, diagrams, and other documentation. They should develop a safety briefing sheet that can be passed to the staging manager for multiagency responses, longer-duration incidents, and unusual fires. Safety briefings are routine for wildland-trained firefighters. A good crew boss repeats safety messages throughout the incident. This behavior should be encouraged at every chance.

Taking the time to diagram and document hazard and safety issues is well received if (or when) the incident is passed to the IMT's ISO.

UNIQUE CONSIDERATIONS AT WILDLAND FIRES

Interface with Aircraft

At a typical Type-5 or -4 fire, you rarely see helicopters and fixed-winged aircraft used for water and retardant drops although more fire departments are utilizing this valuable resource on a quick-call basis, especially during fire season. Being familiar with the hazards associated with these resources is essential to the ISO (**Figure 14-4**). If you have never had training on interfacing with aircraft, seek it out (contractors are usually quite willing during the off-season). You can also access aircraft interface training online at http://iat.nifc.gov.

Inexperienced firefighters or crews working a wildfire may seek out an opportunity to get "slimed" by a fire retardant drop from aircraft; the mind-set is like the warped mentality of having a melted helmet (pun intended!). The accuracy of these drops can be quite exact, but exactness is measured in terms of hundreds of feet. Try to discourage sliming when you hear talk; injuries are likely. However, do not confuse this kind of talk with the message that firefighters are trapped (of course, we prefer to say out of such situations proactively). The trapped firefighters will welcome the drop and accept the risk of a slime injury versus being overrun by fire.

Figure 14-4 *The use of aircraft at wildfires introduces unique hazards. If you do not have training in interfacing with aircraft, seek it out!*

Incident Escalation

Once a fire becomes a Type-3, the ISO function may be transferred to a regional or state person who is trained for and experienced in the role. Having good notes that include the chronological order of events helps the transition. During the transition, a series of briefings and communication is likely; be present and offer your observations to bring the IMT up to speed. The incoming ISO may choose to use you in an ASO role, or you may be simply assigned to the resource pool. The IMT is likely to bring a focus to the incident that may be different from yours, but remember that the goal is the same.

SUMMARY

Wildland fires can include cultivated fields, interface (threatened structures), or wild vegetation. The structural firefighter is likely to be the initial responder to these fires. Judging the likelihood of firefighters being overrun by the fire is the best starting place for the ISO. The ISO should assess the magnitude of the event and classify it based on the eventual resource demand of the fire. There are basically four incident types, with Type-5 being local and Type-1 being a large campaign-like fire. Knowing wildland fire behavior factors and hostile events helps the ISO make overrun judgments.

General ISO duties apply at the wildland fire: monitoring risk, operational effectiveness, and personal safety systems. Trained medical attendants in rehab should carefully monitor the medical status of firefighters. Hydration and electrolyte replacement should commence early. Traffic issues are usually complicated by smoke obscuration. Certain conditions should trigger the assignment of ASOs to help with the ISO functions.

Judging the pace of working crews is essential to managing risk at the wildland fire, especially for structure protection. The principal hazard at most wildland environments includes rapid fire spread, traffic issues, or physical exertion. The ISO should listen to responders and judge whether LCES is being applied and reinforced at the crew level. Questions related to managing time, such as the advent of bad weather or darkness, can help the ISO forecast potential problems. Working crews who have a sense of focus and who display energy show the ISO that rehab efforts are effective. Keeping good documentation can help the ISO hand off critical information when the incident escalates. The interface with aircraft for fire control can lead to concerning issues; seek training.

KEY TERMS

Blow-up A wildland fire term used to describe the sudden advancement and increase in fire intensity due to wind, prewarmed fuels, or topographic features, such as a narrow canyon or a chimney.

Flaring A sudden, short-lived rise in fire intensity, attributed to wind, fuel, or topographical changes. Flaring can be a warning sign of an upcoming blow-up

LCES An acronym that stands for *l*ookouts, *c*ommunication methods, *e*scape routes, and *s*afety zones.

POINTS TO PONDER

Electrocution Hazard at Wildland Fires

Electrical hazards are among the various hazards firefighters face during wildland fire suppression activities. Firefighters performing fireground operations near downed power lines may be exposed to electric shock hazards through the following means:

- Electrical currents that flow through the ground and extend several feet (ground gradient)
- Contact with downed power lines that are still energized
- Overhead power lines that fall onto and energize conductive equipment and materials located on the fireground
- Smoke that becomes charged and conducts electrical current
- Solid-stream water applications on or around energized, downed power lines, or equipment

Let's look at two cases that illustrate these hazards.

Case 1

On June 23, 1999, a 20-year-old male volunteer firefighter was electrocuted while fighting a grass fire. The volunteer firefighter was one of a crew dispatched to a grass fire where a power line was reported to be down. The volunteer firefighter arrived and immediately helped the deputy chief and a firefighter/paramedic extinguish the fire on the east flank. The volunteer firefighter then walked toward a smoldering pile of brush near the downed power line. As he pulled a charged, 1-inch line over the uneven terrain, he apparently tripped and fell onto the 6,700-volt, downed power line. Other firefighters on the fireground used a nonconductive tool to pull the line from under the victim. He was moved to the street, received cardiopulmonary resuscitation (CPR), and was then taken to a local hospital, where he was pronounced dead.

Case 2

On October 4, 1999, a 20-year-old male volunteer firefighter was electrocuted and two other firefighters were injured when they contacted an energized electric fence while fighting a grass fire. Central dispatch notified the fire department of a fire that was started when a downed power line ignited the surrounding grass. The chief arrived first, followed by Engine 1 and two firefighters. The chief indicated to central dispatch and to the responding firefighters that the electric fence bordering the area was energized by the downed power line. The driver of Engine 1 and the three firefighters crawled underneath the bottom wire of the electric fence. They positioned themselves approximately 50 feet from the downed power line and attacked the primary fire. After the fire was extinguished,

(continued)

(continued)

the three firefighters crawled under the fence a second time. It is believed that, when one of the survivors was crawling on her back under the electric fence, a hook from her bunker coat might have contacted the bottom wire of the fence. It is believed that the other two firefighters were shocked while trying to help the firefighter who was still energized. All three were removed from the energized area, and basic first aid procedures were administered until the ambulance arrived. One of the injured was transported by helicopter to an area hospital, and another was transported by ambulance to the local hospital and later to the burn unit of an area hospital. The third firefighter was pronounced dead on arrival at a local hospital.

NIOSH recommends that fire departments take the following precautions with regard to electrical hazards at wildland fires:

- Keep firefighters a minimum distance away from downed power lines until the line is deenergized. The minimum distance is equal to the span between two poles.
- Ensure that the incident commander conveys strategic decisions related to power line location to all suppression crews on the fireground and continually reevaluates fire conditions.
- Establish, implement, and enforce standard operating procedures (SOPs) that address the safety of firefighters when they work near downed power lines or energized electrical equipment. For example, assign one of the fireground personnel to serve as a spotter to ensure that the location of the downed line is communicated to all fireground personnel.
- Do not apply solid-stream water applications on or around energized, downed power lines or equipment.
- Ensure that protective shields, barriers, or alerting techniques are used to protect fire fighters from electrical hazards and energized areas. For example, rope off the energized area.
- Train firefighters in safety-related work practices when working around electrical energy. For example, treat all downed power lines as energized and make firefighters aware of hazards related to ground gradients.
- Ensure that firefighters are equipped with the proper personal protective equipment and that it is maintained in good condition.
- Ensure that rubber gloves and dielectric overshoes and tools (insulated sticks and cable cutters) for handling energized equipment are used by properly trained and qualified personnel.

For Discussion:

1. What are some of the specific similarities in the two cases that led to the fatalities?

2. What are some the obstacles that may interfere with implementing the NIOSH precautions? (Be specific.)

3. The NIOSH precautions are listed for fire departments. What specific precautions should individual firefighters employ to prevent electrocution?

Note:

This case study is extracted from NIOSH Hazard ID #15, *Firefighters Exposed to Electrical Hazards During Wildland Fire Operations.* 2002. Available at: http://www.cdc.gov/niosh/hid15.html.

REVIEW QUESTIONS

1. List and describe five incident types that can be applied to wildland fires.

2. List the three factors that influence fire spread.

3. Define blow-up and flaring.

4. List the leading stresses requiring rehab at the wildland fire.

5. List four situations that may require the appointment of an ASO at wildland fires.

6. List the three most common principal hazards at a wildland fire.

7. Define LCES.

8. Describe the types of behaviors that would indicate rehab efforts are effective.

9. Discuss a troubling issue that may arise when ground firefighter interface with aircraft.

ADDITIONAL RESOURCES

Fire Operations in the Urban Interface. S-205 course, NFES 2171. Boise, ID: National Fire Interagency Fire Center, updated periodically.

Intermediate Wildland Fire Behavior. S-290 course, NFES 2387. Boise, ID: National Fire Interagency Fire Center, updated periodically.

Lowe, Joseph. *Wildland Firefighting Practices.* Clifton Park, NY: Delmar, a division of Thompson Learning, 2001.

National Fire Training. Wildfire training resources: Available at: http://www.national firetraining.net.

Chapter

15

THE ISO AT HAZMAT INCIDENTS

Learning Objectives

Upon completion of this chapter, you should be able to:

- List the federal regulations that may have an impact on ISO functions at hazmat incidents.
- Define the reporting structure for an ASO-HM at a hazmat tech-level incident.
- Define the two overriding risks that the ISO must evaluate at hazmat incidents.
- List the four control zones that need to be established at tech-level hazmat incidents.
- List the three hazmat rehab components that require close evaluation.
- List the ten federal-required components of a hazmat response site safety plan and five hazmat ancillary plans that may require ISO sign-off.
- List five or more alarming hazards at a clandestine drug lab incident.
- List and describe the three strategic goals for the safety section at a WMD/terrorist incident.

INTRODUCTION

Hazardous materials incidents create an enormous potential for short- and long-term disruption of our public areas, infrastructure, and environment (groundwater, habitats, farms, and other resources). Environmental consciousness has driven society to develop specific goals, procedures, and laws to govern hazardous materials production, distribution, and use. The realization of terrorist threats and the use of weapons of mass destruction (WMD) have further heightened social awareness and expectations for proactive threat response and mitigation. The production and distribution of illegal drugs and substances have spawned an epidemic of dangerous drug labs and indiscriminate toxic waste disposal. As public awareness of hazmat issues increase, so do the expectations for fire departments to properly handle hazmat incidents. Given all this, hazmat incidents have become the most regulated of all the incidents to which fire departments might respond. Regulations, the threat of acute and chronic health issues, and the potential damage to the environment have led fire departments to develop a hazmat response system that includes procedures, equipment, and training to help protect firefighters and the community they serve. The appointment of an ISO at a hazmat incident is an integral part of this response system. In fact, the assignment of an ISO at a hazmat technician-level incident is not discretionary; it is mandated by law.[1] As a starting place, the ISO needs to be aware of the many federal regulations that impact the role of the ISO at hazmat incidents:

■ **Note**
In fact, the assignment of an ISO at a hazmat technician-level incident is not discretionary; it is mandated by law.

29 CFR 1910.95, *Occupational Noise Exposure Limits*

29 CFR 1910.120, *Hazardous Waste Operations & Emergency Response Solutions*

29 CFR 1910.134, *Respiratory Protection*

29 CFR 1910.146, *Permit-Required Confined Spaces*

29 CFR 1910.1030, *Blood-Borne Pathogens*

29 CFR 1910.1200, *Hazard Communication*

Where there is regulation, there is liability. These two challenges put tremendous pressure on the fire department—and on the ISO. At the hazmat incident, the ISO should have the professional competencies for the level of incident involved. These competencies are defined in NFPA 472, *Standard for Professional Competence for Responders to Hazardous Materials Incidents*. The standard outlines awareness, operations, command, and technician-level training requirements. For example, if a fire department offers technician-level response, the ISO (by law *and* to be effective) needs to have that level of competency. Where the ISO does not have a technician-level certification, an assistant safety officer—hazmat (ASO-HM) should be appointed to help with technician safety functions. An **assistant safety officer—hazmat (ASO-HM)** is defined as a person who meets or exceeds the NFPA 472 requirements for *Hazardous Materials Technician* and is trained in the responsibilities of the ISO position as it relates to hazmat response. When an ASO-HM is appointed, the ISO retains an overhead safety function responsibility while the ASO-HM works with the technician-level group or branch. Even still, the overhead ISO needs to have the competencies outlined for hazmat command level.

Efforts are underway to standardize language and titles. Still, the ASO-HM may be titled differently based on how the incident management system has been scaled. Some of these titles are:

- Hazmat safety officer
- Hazmat group safety officer
- Hazmat branch safety officer

In this book, the designation ASO-HM is used to indicate the person fulfilling safety functions for the technician-level components at an incident. Organizationally, the ASO-HM should report to and work with the ISO. In reality, the ASO-HM works with three or more people: the ISO, the hazmat branch director (or hazmat group supervisor), and any technical specialists or industry representatives that the plans section chief (or IC) has assigned to assist (**Figure 15-1**).

The purpose of this chapter is to address specific ISO/ASO-HM challenges that need to be considered in performing the assigned safety functions at the hazmat incident. This chapter is *not* designed to replace hazmat-technician professional competencies. The intent is to describe the general duties of the ISO, followed by the application of the ISO Action Model and unique considerations for hazmat incidents. In this discussion, it is assumed that an ISO is in an overhead role and an ASO-HM is assigned for technician-level components.

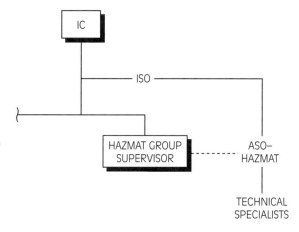

Figure 15-1 *The ASO-HM may actually be working with three or more persons.*

ISO GENERAL DUTIES AT THE HAZMAT INCIDENT

The ISO at a hazmat incident should be intimately familiar with the practices outlined in NFPA 471: *Recommended Practices for Responding to Hazardous Materials Incidents.* The following discussion is based on those recommendations and applies them from the ISO point of view.

Upon arrival and assignment at a hazmat incident, the ISO should ascertain that the initial zone and isolation efforts are in place and effective. From then on, the ISO must take a *strategic* approach, which involves interfacing with and getting input from the other command staff members. The appointment and empowerment of an ASO-HM (and, if needed, other ASOs) help with tactical safety functions. The best place to maintain a strategic profile is at the command post. Occasionally, the ISO may have to rove from the command post to meet with the ASO-HM, the plans section chief (where delegated), and any ASOs.

Monitoring Issues at Hazmat Incidents

Risk   The two overriding risk issues at the hazmat incident are liability and risk communication.

Liability (the legal responsibility) is tied directly to the training levels of responders and further stabilization (entry) objectives. The basic risk liability question is whether a separate, contracted environmental clean-up team can achieve the same entry objectives. Sometimes a fire department hazmat team has to make an entry to verify that the incident is stabilized and savable victims are removed and decontaminated. Once it is decided that a fire department hazmat team entry is warranted, the risk issues associated with hazmat incidents are usually tied to risk communication.

Most firefighters understand that the hazmat incident requires a slow, calculated approach and are likely to back off on risk taking, but others may not understand this need or practice it. The ISO may have to communicate established risk guidelines to other emergency response affiliates or industry representatives at the incident. Some may not understand the magnitude of the event or the potential of the chemicals involved. Industry representatives who have cost responsibility for the incident may front a "no big deal" approach to minimize media coverage or implied liability. Like the IC, the ISO should seek input on the worst-case scenario from the ASO-HM and third-party experts who have no stake in the incident.

Operational Effectiveness   Judging operational effectiveness may be difficult if the ISO is maintaining a strategic profile at the command post. The ISO must therefore rely on the ASO-HM to judge the effectiveness of the technicians' operation. Other assigned ASOs can help with effectiveness judgments of support activities for those not directly involved with the technicians. Most hazmat operations are centered on an action plan that is created, discussed, and communicated *prior* to actual operations. This helps preplan operational effectiveness. The ASO-HM can troubleshoot the action plan to make sure the operation will be effective and to address contingencies should an unplanned event occur.

Personal Safety System Issues at Hazmat Incidents

Accountability Systems   Two systems can be at play: one for the hazmat team members and one for support responders. Cross-communication between the two needs to be encouraged. Realistically, the ISO should deal with strategic accountability and ASOs with tactical accountability.

Control Zones   Two issues apply to control zones at the hazmat incident: language and appropriate PPE.

The typical "hot-, warm-, and cold-zone" language is open to interpretation and may not adequately define the level of PPE required to be in a given zone. As suggested in Chapter 13, the terms "IDLH," "no-entry" (including collapse zone), and "support zone" still apply to hazmat incidents—especially for first responders attempting to zone and isolate the incident. Once a technician-level operation has been developed to further stabilize the incident, however, an additional zone must be defined: the contamination reduction zone. A **contamination reduction zone** is an area where decontamination takes place and includes a safe refuge area for contaminated victims and responders who have left (or rapidly escaped) the IDLH zone. Marking these zones should come with input from the ASO-HM (**Figure 15-2**).

contamination reduction zone
an area where decontamination takes place and includes a safe refuge area for contaminated victims and responders who have left (or who have rapidly escaped) the IDLH zone

Figure 15-2
Developing and marking zones require input from the ASO-HM.

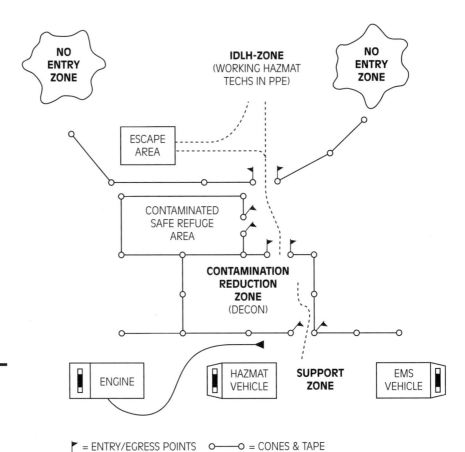

Figure 15-3 *Simple diagrams can help responders understand zone areas and travel paths.*

Other components of effective zoning include travel paths and gateways between zones. Drawing a simple diagram can help responders and entry teams visualize them (**Figure 15-3**).

Additionally, the ASO-HM should take the lead to verify that the level of PPE required for responders working in each zone is appropriate. Persons who transition from one zone to another should follow a prescribed pathway. Prior to leaving the contamination reduction zone, persons may need to be checked with instrumentation in an attempt to verify that the contamination has been decreased.

Radio Transmissions   Technological developments have minimized some communication challenges presented by technician-level operations. However, several issues still exist and some have even been created by the technology. Hazmat protective ensembles may not allow for effective radio communications, or they may introduce multiple radio types and frequencies that need to be monitored. In addition to general radio monitoring, the ISO and ASO-HM must see that backup communication systems are in place. Hand signals, message boards, and tag-line signals can be used as backup systems. While these communication systems are best preplanned and practiced, the ISO/ASO-HM team may suggest developing a spontaneous system on-scene to address unique communication issues. When spontaneous communications systems are developed, they need to be explained to affected incident personnel prior to use.

Rehab   The potential duration of a hazmat operation may require rehab components to span hours or even days. While all the rehab components are applicable, the hazmat incident requires that the ISO/ASO-HM pay particular attention to the evaluation of certain areas:

- *Medical monitoring:* In the structure fire arena, medical monitoring takes place *after* a given assignment has been completed. For hazmat incidents, it is necessary to establish baseline medical monitoring *before* technician stabilization efforts. The stresses created while operating in a totally encapsulating suit can be overwhelming. Having before-and-after responder vital signs can help decision makers make judgments and adjustments regarding working times, rest periods, and active cooling strategies.

- *Sanitation needs:* Extended operations can create sanitation concerns, which are presumably addressed by personnel assigned to the logistics function. These concerns may include the disposal of human waste, the removal of garbage and/or recycle materials, and the disposition of contaminated clothing or equipment. The best hazard mitigation approach in dealing with these issues is *separation.*

- *Food service:* As wrong as it sounds, feeding firefighters at a structure fire rarely addresses issues of distance and cleanliness. At the hazmat incident, these two issues *must* be addressed. Food storage, preparation (if any), and distribution should be well clear of working areas. The food service area should include areas for additional hand and face washing (further decon!).

Defining Other Needs at Hazmat Incidents

Traffic   The location of the hazmat incident is likely to create traffic conditions that could harm firefighters. In addition to roadway traffic, the ISO must consider safety issues associated with railway, air, and waterway traffic. The basic approach to all traffic issues is simply to get rid of them: Too many variables exist that should not be left to chance.

A less obvious issue is people traffic. Defining specific shuttle pathways, escape zones, and zone transition gateways can pay safety dividends.

Need for ISO Assistance   In addition to the assignment of an ASO-HM and one or more ASOs, the incident commander may request technical specialists, corporate risk managers, process experts, and public health representatives to respond and assist with specific planning functions. These reps can also provide safety-specific information for the ISO. If responders have been exposed to chemicals, the ISO should consider calling the department HSO or infection control officer to help with procedural issues and documentation for the exposure.

APPLYING THE ISO ACTION MODEL AT HAZMAT INCIDENTS

The ISO Action Model is applicable to hazmat incidents.

Risk Evaluation at the Hazmat Incident

The principle of "risk a life to save a life" may not be appropriate for the hazmat incident. The ISO and ASO-HM must strive to agree on the overall risk profile of the incident, and the ISO may need to communicate an acceptable risk profile for nonfire service personnel involved with the incident.

When evaluating the pace of the hazmat incident, the ISO should adopt the time-tested phrase: "If you don't know, don't go, 'cause it might blow." A slow, methodical, and intellectual approach is the best pace for hazmat incidents (**Figure 15-4**). Assigned ASOs and the ASO-HM should watch for crews or individual responders who are out of synch with the incident pace; they are taking an inappropriate risk.

Recon Evaluation at the Hazmat Incident

Upon arrival and assignment at a hazmat incident, the ISO should confirm that initial responders have appropriately zoned and isolated the hazard. The ISO and ASO-HM must verify that the defined zones and gateways are appropriate and well marked. In doing so, the ASO-HM should seek input from the technical reference specialist (when assigned) or from the technical reference library—and not rely on previous experiences. The tech reference specialist can also help the ISO/ASO-HM define the principal hazard, environmental integrity (and threat), and the effects of the surrounding elements and the exposure threats to responders.

! Safety
The ISO and ASO-HM must verify that the defined zones and gateways are appropriate and well marked. In doing so, the ASO-HM should seek input from the technical reference specialist (when assigned) or from the technical reference library—and not rely on previous experiences.

Figure 15-4 *A slow, methodical, and intellectual approach is the best pace for hazmat incidents.*

Defining the Principal Hazard   Typically, the chemical involved dictates the principal hazard. When defining the principal hazard, the ISO/ASO-HM must understand the potential for detonation, ground leaching, gas plume spread, and liquid flow. The hazmat tech team likely has this information, and it should be passed to the IC/ISO for review.

Defining Environmental Integrity   Environmental integrity considerations at the hazmat incident typically include the weather, infrastructure stability, container condition, and hazardous energy. Assigning a value to the integrity of each of these factors is essential for the ISO/ASO-HM team:

- Stable—not likely to change
- Stable—may change
- Unstable—may require attention
- Unstable—requires immediate attention

Defining Physical Surroundings   The physical location of the hazmat incident helps the ISO define the impact of the surroundings. For example, if the incident is taking place on a public roadway, features such as terrain, foliage, curbs, posts, fences, drainages, barriers, and population density need to be evaluated. Conversely, an incident at a fixed processing or distribution facility presents physical hazards such as the presence of complex equipment, access restrictions, security features, and hazardous energy.

Crew Exposure to Hazards   The equation for expressing crew exposure changes a bit for the hazmat incident. The relationship can be expressed as:

Physical hazards + Chemical properties + Crew mitigation efforts
= Crew hazard exposure

The ASO-HM is in the best position to evaluate issues regarding tools, tasks, teams, and rapid withdrawal factors. Rapid intervention is far from rapid at the hazmat incident. Plan for the worst, but ensure that clear direction is established for the *activation* of the rapid intervention crew. Pay attention to escape paths, safe refuge areas for contaminated victims and responders, and zone gateway control.

Resource Evaluation at the Hazmat Incident

Many fire departments use interagency agreements to assemble enough hazmat technicians to pull off a mitigation action plan. The assembly of interagency teams takes time. From the ISO/ASO-HM perspective, the mitigation effort should not commence until adequate teams are assembled, equipped, and briefed. Obviously, the history of working together that these interagency teams have is important: Do they train together often? Are there equipment compatibility issues? Do all teams use the same IMS structure and language? Answers to these questions can help the ISO make judgments about the potential for positive or negative outcomes.

Time   For structure fires, the ISO needs to project on-scene time. Doing so may not be practical at the hazmat incident due to its size, complexity, and slow pace. The ISO should, however, manage the impacts of time passage. Reflex time for any unplanned event is delayed—and should be—so that ramifications can be studied.

Personnel   Of particular concern is the training of personnel to perform their assigned task. Responders who are assigned tasks that exceed their training can create an injury potential, along with liability should an injury occur. When a person's level of training is not known or confirmed, the ISO/ASO-HM should relay the concern to the hazmat group supervisor and/or IC.

Equipment   Understanding the types of equipment present, what is needed, and what is still coming can help the ASO-HM evaluate adequacy. At times, specialized equipment may be required to stabilize the hazard. If the hazmat team has never worked with the specialized equipment, the ASO-HM may need to encourage additional on-the-spot training time to ensure that the technicians can operate the equipment in full encapsulating ensembles.

Operations taking place well within a building (over 300 feet) present unique concerns regarding air use, equipment shuttling, lag time, and rapid egress. The same can be said for outdoor hazmat incidents with a large IDLH or numerous no-entry zones. The use of golf carts or other shuttle vehicles can maximize air usage and minimize escape times.

Report Issues at the Hazmat Incident

■ Note

A tech-level stabilization effort requires the **formal** *development and delivery of a site safety plan and safety briefings. In this context, "formal" means documented.*

Of all the ISO Action Model components, reporting requires the greatest effort and commitment from the ISO. A tech-level stabilization effort requires the *formal* development and delivery of a site safety plan and safety briefings. In this context, "formal" means documented. The IC is tasked with developing a written site safety plan and typically delegates this to the ISO. Federal law specifically requires a written site safety plan that includes the following elements:[2]

- Safety, health, and hazard risk analysis, including entry objectives
- Site organization, including the training and qualifications of responders
- Identification of the exact type of PPE required for the tasks performed by responders
- Medical monitoring procedures
- Environmental monitoring and sampling procedures
- Site control measures, including exact control zone locations and gateway marking
- Decontamination procedures
- Predefined responder emergency plans (for fires, medical emergencies, and rapid intervention)
- Confined space entry procedures, including intervention and escape plans
- Spill containment procedures, including container-handling measures.

Documenting that incident responders have received appropriate safety briefings is included in the site safety plan requirement. Additionally, the ISO or ASO-HM may have to sign off on other developed plans (see Hazmat Incident Plans That May Require ISO Sign-Off).

The 15-minute rule for IC face-to-face communication is impractical at the hazmat incident. The ISO should confer with the IC to establish a time cycle or benchmark triggers for face-to-face updates. The ISO is likely

---

Hazmat Incident Plans That May Require ISO Sign-Off

In addition to the site safety plan, the ISO and/or ASO-HM may have to sign off on numerous plans that have been developed for the hazmat incident:

- Incident action plan
- Communications plan
- Exposure protection plan
- Evacuation plan
- Mass-victim decontamination plan
- Responder decontamination plan
- Traffic plan
- Medical plan
- Demobilization plan
- Other site-specific plans

■ Note

Hazmat documentation is not subject to any statute of limitations and may be used for litigation purposes years—or even decades—after the incident. Be thorough!

to be required to fill out some sort of unit log to document his or her efforts and actions during the incident. Be sure to set aside time to document all requirements as the incident evolves and finally concludes. Hazmat documentation is not subject to any statute of limitations and may be used for litigation purposes years—or even decades—after the incident. Be thorough!

UNIQUE CONSIDERATIONS AT THE HAZMAT INCIDENT

Clandestine Drug Labs

The response of a fire department to a fire, explosion, or EMS call at a suspected or confirmed clandestine drug lab and/or waste disposal site taxes even the most experienced responder. Drug labs and waste sites can be found in homes, businesses, hotel rooms, and vehicles of all shapes and sizes. A law enforcement raid on a suspected drug lab may call for the standby of fire personnel and hazmat team responders. Firefighters may discover the remains of an active or abandoned drug lab in the course of other incident responses or activities. In either case, the responding ISO/ASO-HM has to include additional considerations in evaluating hazard safety.

The response to a clandestine drug lab or waste site typically involves a multiagency effort that includes law enforcement, EMS, fire, public health, environmental protection regulators, and even social services. At any time, the primary management authority of the incident may change. These events must be preplanned, and the ISO must understand who has primary control authority for any phase of the incident. Other than fire control and victim treatment, the law enforcement agency usually has primary control responsibilities for criminal investigative purposes.

The suspected clandestine lab presents alarming hazards to fire department personnel who are actively involved with fire control or stabilization activities. These include:

- Poor ventilation

- Flammable/toxic atmospheres

- Incompatible chemicals

- Chemical reactions in progress

- Unidentified chemicals and/or containers

- Unstable and/or leaking containers

- Booby traps (improvised firearms, incendiary/explosive devices, and other such devices)

When a suspected drug lab is discovered as part of a fire, odor investigation, EMS, or other nonhazmat response, the fire department should immediately notify law enforcement and tech-level hazmat teams. The ISO should meet with the incident commander and offer solutions to initiate a careful (but immediate) withdrawal of responders. Chemical exposure, decontamination, isolation procedures, and evidence protection become priorities. Routine firefighting tasks, such as utility control, overhaul, and debris removal, should not be attempted unless directed by law enforcement or forensic chemists familiar with clandestine drug lab intricacies.

● **Caution**

Routine firefighting tasks, such as utility control, overhaul, and debris removal, should not be attempted unless directed by law enforcement or forensic chemists familiar with clandestine drug lab intricacies.

Weapons of Mass Destruction

Once an incident has been classified as a suspected terrorist event using a weapon of mass destruction (WMD), the FBI takes the lead, as provided by Presidential Decision Directive Thirty-Nine (PDD 39). The local ISO is usually replaced by an overhead safety officer who responds as part of an incident management team (IMT) following a terrorist attack. But the transition time between the onset of the suspected terrorist WMD event and the pass-off to an IMT can take time, lasting from a few hours to several days depending on the magnitude, geographical location, and competing demands of other simultaneous events. Developing a local WMD plan that also addresses ISO functions assists in implementing an organized approach to the crime and reduces responder exposure and mortality. The ISO must remain *strategic* and

■ **Note**

Developing a local WMD plan that also addresses ISO functions assists in implementing an organized approach to the crime and reduces responder exposure and mortality.

■ **Note**

In brief, the ISO should coordinate a *quick-in/quick-out* approach for immediate rescues, then support a *back-off* posture. Victims and exposed firefighters should be isolated until *clean* or *contaminated* determinations can be made. Staged equipment should be out of the sight of gathered spectators because the terrorist may be among them and waiting for an opportunity to compound the event by attacking the responders.

focus on the basic safety issues associated with the Office of Domestic Preparedness Emergency Responder Guidelines:

- Recognition
- Detection
- Self-protection
- Crime scene preservation
- Scene security
- Notifications

In brief, the ISO should coordinate a *quick-in/quick-out* approach for immediate rescues, then support a *back-off* posture. Victims and exposed firefighters should be isolated until *clean* or *contaminated* determinations can be made. Staged equipment should be out of the sight of gathered spectators because the terrorist may be among them and waiting for an opportunity to compound the event by attacking the responders. If it is not possible to stage out of sight, then security measures are warranted (up to and including armed guards!).

Once the initial overall strategic concerns are addressed, the ISO should move to create an integrated approach to incidentwide safety. The ISO should recognize that the WMD incident will become a multiagency event and will likely require the ASOs to deal with the high pressure and time constraints of the postdisaster environment. The strategic goals of the ISO and ASOs should be:

■ **Note**

The ISO should recognize that the WMD incident will become a multiagency event and will likely require the ASOs to deal with the high pressure and time constraints of the postdisaster environment.

- *Gather RECON and threat information:* Use ASOs and representatives from law enforcement and other emergency response agencies to ascertain the threat potential, resource status, chemical agent type, and possible acute or chronic health concerns. This information collection will likely be accomplished by the plans section; the point is that the ISO and ASOs have the information.

- *Analyze options:* Draw from the technical expertise of multiple responding organizations and lean toward the worst case. Remember that worst-case analysis is for planning and not for public information. Protect this analysis to prevent panic.

- *Develop a safety action plan:* The action plan should focus on sustainability measures to protect the health of responders across organizational boundaries. Ancillary agencies may not be familiar with the safety officer function; communication in a simple, yet compelling manner helps nonfire service responders understand and comply with safety plans. Accountability, PPE, rehab, and zoning issues are a great starting place for the safety action plan. Once the plan has been delivered to responders, the ISO should focus on expanding his or her role into manageable parts (see Chapter 11), and begin addressing safety and health issues prior to the arrival of an IMT (**Figure 15-5**).

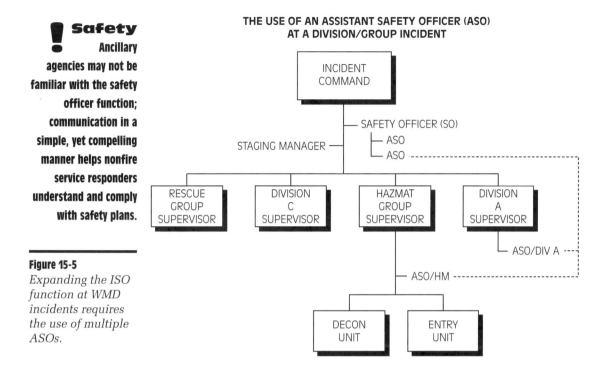

**THE USE OF AN ASSISTANT SAFETY OFFICER (ASO)
AT A DIVISION/GROUP INCIDENT**

> ❗ **Safety**
>
> **Ancillary agencies may not be familiar with the safety officer function; communication in a simple, yet compelling manner helps nonfire service responders understand and comply with safety plans.**

Figure 15-5
Expanding the ISO function at WMD incidents requires the use of multiple ASOs.

SUMMARY

Hazardous materials incidents require specialized training (and certification) due to their complex and libelous nature. The actions (and inactions) of hazmat responders may have significant acute (short-term) and chronic (long-term) ramifications on the environment and on public health. As a result, hazmat incidents have become one of the most regulated of all the incident types that an ISO may face.

The use of an ASO-HM who works for an overhead ISO is perhaps the best way to address the intricacies of the hazmat incident. The ISO can address general ISO duties (with input from the ASO-HM), while the ASO-HM focuses on technician-level issues.

Risk issues include proper training for hazmat support assignments and the communication of hazards to nonfire service responders.

Control zones should be clearly communicated and include IDLH, no-entry, contamination reduction, and support. Rehab evaluation should include particular attention to medical monitoring before and after entry, sanitation concerns, and food service. Remember that separation (distance) is a key concept for sanitation and food service areas.

The ISO Action Model can be easily applied to a hazmat incident. The report component requires significant effort because of the mandated need for developing and delivering formal (documented) plans. A site safety plan must be developed by the IC, who typically delegates the requirement to the ISO. There are ten essential components of the site safety plan.

Suspected clandestine drug labs and weapons of mass destruction (WMD) incidents can stress responders and the ISO. Having a defined plan that emphasizes minimal exposure works for both drug labs and WMD incidents. At WMD incidents, the ISO must remain strategic and initiate an integrated, incidentwide safety management approach prior to the arrival of an IMT.

KEY TERMS

Assistant safety officer—hazmat (ASO-HM) A person who meets or exceeds the NFPA 472 requirements for Hazardous Materials Technician and is trained in the responsibilities of the ISO position as it relates to hazmat response.

Contamination reduction zone An area where decontamination takes place and includes a safe refuge area for contaminated victims and responders who have left (or who have rapidly escaped) the IDLH zone.

POINTS TO PONDER

The School Bus Fire

The fire department was dispatched to a report of a vehicle fire with structures threatened. Upon arrival, the first-due engine found an old school bus with fire blowing out its rear portion. The bus appeared to have been converted to a recreational vehicle with an LP tank strapped to the outside. Five feet from the bus was a small detached garage that was beginning to ignite. Two lines were deployed to attack the fire and keep the LP tank cool. Once the fire was knocked down from the outside, one attack team moved to the front bus access door to extinguish the remaining fire. Upon entry, they found an unresponsive victim of the smoke. The victim was removed and passed to ambulance personnel.

During the attack and rescue, the IC and ISO began their respective duties. At some point, a later-arriving sheriff deputy approached the IC and said that the bus was a suspected clandestine drug lab, according to an undercover drug agent who was in the area. The sheriff went on to explain that one of the bystanders (who was dressed like a vagrant) didn't want to break "cover" and was actually on surveillance at a nearby home when the fire was discovered.

The IC immediately relayed this info to the ISO and the two agreed to order a hazmat team response. Crews were withdrawn, and the department's clandestine drug lab protocol was implemented.

For Discussion:

1. Generally speaking, what issues would you expect to arise when the incident transitions from a fire attack to a hazmat incident?

2. As the IC, would you let the crews finish extinguishing the fire? If so, what restrictions would you communicate to the crews?

3. Whom would you consider to be "contaminated" at this incident? What would you do for those contaminated?

4. As an ISO, what types of things would you do once you found out that a clandestine drug lab was suspected?

REVIEW QUESTIONS

1. List the federal regulations that may have an impact on ISO functions.

2. To whom does the ASO-HM report at a hazmat tech-level incident?

3. With whom does the ASO-HM likely work at a typical hazmat tech-level incident?

4. Persons not trained for their hazmat incident assignments create two risks. What are they?

5. What is a contamination reduction zone and where is it located?

6. List the three hazmat rehab components that require close evaluation.

7. What are the ten federally required components of a hazmat response site safety plan?

8. List five hazmat ancillary plans that may require ISO sign-off.

9. List five or more alarming hazards at a clandestine drug lab incident.

10. List and describe the three strategic goals for the ISO at a WMD/terrorist incident.

ADDITIONAL RESOURCES

Bevelacqua, Armando S., and Richard H. Stilp. *Hazardous Materials Field Guide.* Clifton Park, NY: Delmar, a division of Thomson Learning, 1998.

Bevelacqua, Armando S., and Richard H. Stilp. *Terrorism Handbook for Operational Responders,* 2nd ed. Clifton Park, NY: Delmar, a division of Thomson Learning, 2004.

Buck, George. *Preparing for Terrorism: An Emergency Services Guide.* Clifton Park, NY: Delmar, a division of Thomson Learning, 1998.

Hawley, Christopher. *Hazardous Materials Incidents.* Clifton Park, NY: Delmar, a division of Thomson Learning, 2004.

NFPA 471: *Recommended Practices for Responding to Hazardous Materials Incidents.* Quincy, MA: National Fire Protection Association, 2002.

NFPA 472: *Standard for Professional Competence for Responders to Hazardous Materials Incidents.* Quincy, MA: National Fire Protection Association, 2002.

Schnepp, Rob, and Paul W. Gantt. *Hazardous Materials, Regulations, Response, and Site Operations.* Clifton Park, NY: Delmar, a division of Thomson Learning, 1999.

"Safety Management in Disaster and Terrorism Response." *Protecting Emergency Responders,* Vol. 3, NIOSH Publication No. 2004-144. Available at: www.cdc.gov/niosh/docs/2004-144/chap7.html.

NOTES

1. 29 CFR 1910.120(q)(3)(G) requires the use of a safety officer at hazmat incidents.

2. Paraphrased from 29 CFR 1910.120.

Chapter

16

THE ISO AT TECHNICAL RESCUE INCIDENTS

Learning Objectives

Upon completion of this chapter, you should be able to:

■ List several regulations that outline response requirements for tech-rescue incidents.

■ Name the incidents that require or benefit from the assignment of an ASO-TR.

■ Describe the IMS organizational relationship of an ASO-RT at tech-rescue incidents.

■ List the two rehab issues that require special attention at tech rescue incidents and describe the "on-deck" system for crew rotation.

■ Name the four ways to classify a building collapse.

■ List five hazards associated with industrial entrapments.

■ Define "LCES" and how it can be used at a cave-in incident.

■ List six hazards associated with water rescues.

■ List five hazards associated with high-angle rescues.

■ Name five circumstances where a duty ISO should implement a discretionary response to motor vehicle accidents and diagram a strategic approach to protect rescuers at roadway incidents.

■ Discuss the potential hazards and problems that may have an impact on a railway incident and an aircraft incident.

INTRODUCTION

Addressing the safety-related aspects of the various types of technical rescue incidents is like trying to explain the variations in musical genres: There are so many styles and subsets of styles! However, one thing is certain: If someone becomes trapped (and injured), the fire department is called. Regardless of how a victim became trapped or what is causing the entrapment, the fire department is called upon to find a positive solution. Herein lies the trap. Firefighters' "can-do" attitude compels them to action, whether they are trained for the specific situation or not. Resourcefulness, inventiveness, willingness, and compassion drive firefighters to a solution. What a great problem to have! These attributes are what endear firefighters to their communities. From the ISO perspective, the selfsame attributes lead directly to firefighter injuries and deaths that only compound the incident. Firefighting history is full of examples of willing firefighters who got caught up in a situation that soon overwhelmed them, and the result was tragedy. Repeated tragedies have spawned regulations for especially risky rescue incidents. Additionally, fire service members have pushed for specific training standards to prevent reoccurrences.

This chapter presents tangible, broad-based issues for the ISO to consider when performing at the tech-rescue incident. As with hazmat incidents, a tech-rescue incident may fall under federal regulations. Being familiar with these regulations is essential. The following CFRs may have an impact on the functions of the ISO at a tech rescue:

> 29 CFR 1910.95, *Occupational Noise Exposure Limits*
>
> 29 CFR 1910.120, *Hazardous Waste Operations & Emergency Response Solutions*
>
> 29 CFR 1910.134, *Respiratory Protection*
>
> 29 CFR 1910.146, *Permit-Required Confined Spaces*
>
> 29 CFR 1910.147, *The Control of Hazardous Energy (Lockout/Tag-Out)*
>
> 29 CFR 1910.1030, *Blood-Borne Pathogens*
>
> 29 CFR 1910.1200, *Hazard Communication*
>
> 29 CFR 1910.1926, *Excavations, Trenching Operations*

When there is regulation, legal liability exists. These two challenges have led fire departments to develop tech-rescue response systems that address procedures, training, equipment, and command elements. The assignment of an incident safety officer is mandatory for confined space, trench and hazmat incidents. For certain types of rescues, the ISO should have the professional competencies for the level of incident involved. Specifically, building collapse rescue operations require rescue technicians to meet competencies defined in NFPA 1670, *Standard on Operations and Training for Technical*

■ **Note**
Regardless of how a victim became trapped or what is causing the entrapment, the fire department is called upon to find a positive solution.

■ **Note**
Firefighting history is full of examples of willing firefighters who got caught up in a situation that soon overwhelmed them, and the result was tragedy.

■ **Note**
The assignment of an incident safety officer is mandatory for confined space, trench and hazmat incidents.

**assistant safety
officer—rescue tech
(ASO-RT)**
a person who meets
or exceeds the NFPA
1670 requirements for
Rescue Technician and
is trained in the re-
sponsibilities of the
ISO position as it re-
lates to the specific
rescue incident

■ **Note**

**When the ISO does not
have technician-level
training and/or
certification, an
assistant safety
officer—rescue tech
(ASO-RT) should be
appointed to help
with technician safety
functions.**

Search and Rescue Incidents. The standard outlines awareness, operations, and technician-level training requirements. For example, if a fire department offers technician-level response, the ISO (to be effective) needs to have that level of competency. When the ISO does not have technician-level training and/or certification, an assistant safety officer—rescue tech (ASO-RT) should be appointed to help with technician safety functions. An **assistant safety officer—rescue tech (ASO-RT)** is a person who meets or exceeds the NFPA 1670 requirements for *Rescue Technician* and is trained in the responsibilities of the ISO position as it relates to the specific rescue incident. When an ASO-RT is appointed, the ISO retains overall safety function responsibility, while the ASO-RT works with the technician-level group or branch.

As with hazmat, the ASO-RT may be titled differently based on how the incident management system has been scaled. Some of these titles are:

• Rescue safety officer

• Rescue group safety officer

• USAR safety officer (for organized urban search and rescue teams)

In this book, ASO-RT is used to identify the person fulfilling safety functions for the technician-level components at an incident. Organizationally, the ASO-RT should report to and work with the overhead ISO. In reality, the ASO-RT works with three or more people: the ISO, the rescue branch director (or rescue group supervisor), and any technical specialists involved **(Figure 16-1)**.

The purpose of this chapter is to address specific ISO/ASO-RT challenges that need to be considered in performing the assigned safety functions at the tech-rescue incident. This chapter is *not* designed to replace technical rescue professional competencies. We depart from the format used to address general

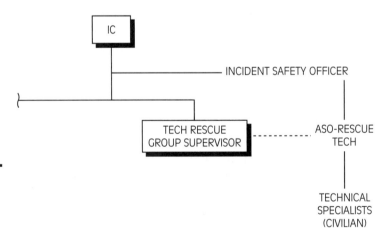

Figure 16-1   *The
ASO-RT actually
works with three or
more persons.*

duties and the application of the ISO Action Model because each type of technical rescue has unique characteristics and one model does not fit all. However, we start with the general ISO duties applicable to all tech rescues and assume that the ISO is in an overhead role and an ASO-RT is assigned for technician-level components.

ISO GENERAL DUTIES AT THE TECH-RESCUE INCIDENT

Because of the vast array of tech-rescue variations, the ISO must gain a strong sense of the situation status (sitstat), which includes victim location, predicament, rescue likelihood, and the integrity of the surrounding environment. Additionally, the ISO must gain an understanding of committed resources (restat) and what is being planned. The best place to get this information and awareness is at the command post. Once in possession of a sitstat and restat, the ISO should shift into a strategic mode and work with the IC, ASO-RT, and technical specialists to develop and troubleshoot action plans. The ISO may have to rove occasionally to meet with the ASO-RT, plans section chief (when delegated), and ASOs.

Monitoring Issues at Tech-Rescue Incidents

Risk   As at a structure fire, the ISO must evaluate the rescue profile of victims who are entrapped or thought to be entrapped. The nature of building collapses is such that the rescue profile is an unknown—a void space may have trapped savable victims. In other cases, the rescue profile of a victim is obvious. When a victim is obviously deceased (or has been determined to be deceased), a recovery mode should be adopted and risks reduced. When the fire department effort switches to a reduced-risk recovery mode, distraught onlookers may attempt to jump in, presenting a difficult situation that can lead to emotional outbursts and expectation stress on responders. When this situation develops, the fire department has to attempt to convince coworkers, relatives, and Good Samaritans that their involvement will only complicate the situation and that their involvement diverts responder attention away from victim efforts. It is not suggested that the ISO be responsible for convincing these onlookers. The ISO needs to be aware of their tendencies and focus on reducing responder stresses that can cause injury.

Operational Effectiveness   It may be difficult to judge operational effectiveness if the ISO is maintaining a strategic profile at the command post. The ISO must rely on the ASO-RT to judge the effectiveness of the technician operation. An additional ASO can help with effectiveness judgments of support responders who are not directly involved with the technicians. The nature of tech

Safety
When the fire department effort switches to a reduced-risk recovery mode, distraught onlookers may attempt to jump in, presenting a difficult situation that can lead to emotional outbursts and expectation stress on responders. When this situation develops, the fire department has to attempt to convince coworkers, relatives, and Good Samaritans that their involvement will only complicate the situation and that their involvement diverts responder attention away from victim efforts.

rescues is such that operational efforts are constantly evolving and shifting. As responders make progress, additional hazards, issues, and complications arise. The ASO-RT and ASOs have to establish close communication with working responders so that operational effectiveness can be continually evaluated. When crews are not achieving desired outcomes, the overall action plan may have to be altered. The ASO-RT and other ASOs must report ineffectiveness to the ISO so that he or she can work with the IC to revise the incident action plan.

Personal Safety System Issues at Tech-Rescue Incidents

■ **Note**
The real accountability issue at tech rescues is the potential for freelancing and self-deployment, as firefighters rush against the clock to save a victim.

Accountability Systems   Tracking assigned resources at the tech rescue should follow the established procedures applicable to other types of incidents. The real accountability issue at tech rescues is the potential for freelancing and self-deployment, as firefighters rush against the clock to save a victim. The ISO and any assigned ASOs need to understand that it is in firefighters' nature to do what they think is best, yet freelancing is inherently unsafe. This is an *issue* that may have to be addressed when risks outweigh benefits.

Control Zones   The differences between IDLH, no-entry, and support zones may be measured in inches at some tech-rescue calls. Communication is essential. The ASO-RT can establish and relay the delineation of zones. ASOs need to monitor the established zones and offer reminders to crews when marking is impractical.

Establishing the required levels of personal protective equipment at the tech rescue often falls on the ISO, who uses information from the ASO-RT to make this judgment. The ISO should establish PPE requirements based on a realistic worst-case scenario—not a popular decision. The PPE requirements may be different for each of the established zones, and ASOs can help monitor compliance.

■ **Note**
To help with clarity, a cheat sheet should be created so that support responders do not miss an important signal.

Radio Transmissions   The nature of some tech-rescue events requires technicians to communicate constant instructions via radio: "A little to the left—hold" . . . "a little to the right—hold" . . . "more slack." These transmissions can tie up radio frequencies. Using small talk-around radios can free up tactical channels, provided their users can also monitor the tac channel. The ASO-RT should have the ability to monitor talk-arounds. Hand signals, message boards, and tag-line signals may also have to be used. Make sure support personnel are clear on the use of these devices; they may have learned to use them or practiced with them years ago. To help with clarity, a cheat sheet should be created so that support responders do not miss an important signal.

Rehab The potential duration of a tech-rescue operation may require rehab components to span hours or even days. While all the rehab components are applicable, the tech-rescue incident requires particular attention in the evaluation of certain areas:

on-deck system

an organized system in which a working team is replaced with another working team that is already dialed in and ready to replace them

- *Perceived comfort:* When a rescue is underway, responders may not admit that they are in need of rehab. Careful monitoring is essential. Do not allow teams to make rehab decisions based on their perceived comfort. Creating a defined work period helps. Like hockey players, responders can be assembled in line changes and rotated on and off task. This approach can be further enhanced if the line changes practice an "on-deck" system. An **on-deck system** is an organized system in which a working team is replaced with another working team that is already dialed in and ready to replace them. Once replaced, the rotated workers go to rehab, while a crew from rehab rotates into the vacated on-deck position. Those in rehab should have an opportunity to disengage mentally. Only when they rotate to the on-deck position should they reengage.

- *Energy replacement:* Most tech-rescue incidents challenge firefighters' concentration and physical stamina. Chapter 10 outlined efficient fueling strategies that can make sure that muscle and brain cells work at optimal levels.

Defining Other Needs at Tech-Rescue Incidents

Traffic A dramatic or unusual rescue attempt undoubtedly attracts media coverage, which increases the numbers of onlookers. Congestion is a real issue that requires attention. While traffic sounds like a law enforcement duty, the ISO needs to consider the impact of the congestion and media coverage. Firefighters who have never "performed" in front of the media and dozens (if not hundreds) of onlookers may become stressed or distracted. ASOs should watch for signs that a firefighter is becoming distracted and offer friendly or humorous concentration reminders.

In addition to roadway congestion, the ISO must be alert for safety hazards associated with railways, air traffic, and waterways. Make sure a travel corridor is maintained for additional resources and equipment shuttling. When rescue helicopters and/or air ambulances are ordered, ensure that an ASO has evaluated landing zones and that the landing zone is separated from the actual rescue location and assembled crowds.

Need for ISO Assistance The incident commander may request technical specialists, risk managers, process experts, and consultants to respond and assist with specific planning functions. These reps can also provide safety-specific information to the ISO. For especially traumatic or emotional incidents, consider implementing critical incident stress management procedures.

CONSIDERATIONS AT SPECIFIC TECH-RESCUE INCIDENTS

Tech-rescue incidents, while diverse, can be classified into several categories that, in some cases, are guided by regulated and response/training standards (**Figure 16-2**). Some may argue that knowing all the information contained in the guiding documents is impossible. The intent therefore is that you are aware of them and have perused their content as part of your professional front-loading. Having the ability to retrieve the critical information in the documents could be useful during the incident, and evolving technology (laptops computers, cell technology, and other devices) can help you with information retrieval.

We now discuss the unique hazards and considerations for the ISO/ASO-RT in each of the tech-rescue categories.

INCIDENT TYPE	GUIDING DOCUMENTS
BUILDING COLLAPSE	• 29 CFR 1910.146, CONFINED SPACES • 29 CFR 1910.147, HAZARDOUS ENERGY • 29 CFR 1910.132, PPE • 29 CFR 1910.1030, BLOODBORNE PATHOGENS • NFPA 1670, TECH RESCUE INCIDENTS
INDUSTRIAL ENTRAPMENT	• 29 CFR 1910.95, NOISE • 29 CFR 1910.120, HAZMAT • 29 CFR 1910.147, HAZARDOUS ENERGY • 29 CFR 1910.132, PPE • 29 CFR 1910.1030, BLOODBORNE PATHOGENS
TRENCH/EARTHEN/MATERIAL CAVE-IN	• 29 CFR 1926.650, TRENCH/COLLAPSE • NFPA 1670, TECH RESCUE INCIDENTS
WATER	• PROFESSIONAL ASSOCIATION OF DIVING INSTRUCTORS (PADI), RESCUE DIVER TRAINING
HIGH ANGLE	• NFPA 1983, ROPES AND HARNESSES
CONFINED SPACE	• 29 CFR 1910.146, CONFINED SPACES • 29 CFR 1910.147, HAZARDOUS ENERGY • 29 CFR 1910.132, PPE • NFPA 1670, TECH RESCUE INCIDENTS
ROADWAY/TRANSPORTATION MODES	• 29 CFR 1910.120, HAZMAT

Figure 16-2 *A sample of technical rescue categories and their associated documents.*

Figure 16-3 *Classifying building collapses is a good starting place for understanding associated hazards. This is a "moderate" collapse. (Photo courtesy of Edwina Davis.)*

basic/surface collapse
collapse in which victims are easily accessible and trapped by surface debris; loads are minimal and easily moved by rescuers; the threat of secondary collapse is minimal

light collapse
a collapse in which usually a light-frame (wood) partition collapses and common fire department equipment (from engine and truck companies) can access or shore areas for search and extrication; the threats of secondary collapse can be mitigated easily

Building Collapse

Classifying building collapse incidents is a good starting place for understanding associated hazards (**Figure 16-3**). While different classification methods exist—and what is suggested here may be overly simple for the rescue technician—the ISO can gain some hazard forecasting ability by classifying the incident (see Building Collapse Classifications).

Building Collapse Classifications

Basic/Surface Collapse   Victims are easily accessible and trapped by surface debris. Loads are minimal and easily moved by rescuers. The threat of secondary collapse is minimal.

Light Collapse   Usually light-frame (wood) partition collapse, and common fire department equipment (from engine and truck companies) can access or shore areas for search and extrication. The threats of a secondary collapse can be mitigated easily.

Moderate Collapse   This is an ordinary construction collapse that involves masonry materials and heavier wood. Lightweight construction with unstable, large, open spans should also be classified as moderate. Significant void space
(continued)

moderate collapse
a collapse of ordinary construction that involves masonry materials and heavy wood; lightweight construction with unstable large open spans should also be classified as moderate; significant void space concerns are present

heavy collapse
a collapse in which stressed concrete, reinforced concrete, and steel girders are impeding access; included are collapses that require the response of USAR teams and specialized equipment and collapses that threaten other structures or that involve the possibility of significant secondary collapse

(*continued*)

concerns are present. Victims may be trapped by load-bearing members requiring heavy lifting equipment. Serious attention should be given to secondary collapse.

Heavy Collapse   Stressed concrete, reinforced concrete, and steel girders are impeding access. These collapses require the response of USAR teams and specialized heavy equipment. Collapses that threaten other structures or significant secondary collapse also fall into this category.

Moderate and heavy collapses should trigger the ISO to request an ASO-RT from the IC if one has not already been appointed. Additional ASOs may also be required, based on the size and complexity of the incident, to address collapse hazards (see Building Collapse Hazards).

Building Collapse Hazards

- Falling/loose debris
- Instability
- Secondary collapse
- Poor air quality/dust
- Unsecured hazardous energy
- Weather exposure
- Blood-borne pathogens
- Difficult access/escape options
- Sharp or rugged debris
- Poor footing

● **Caution**
Drywall/concrete dust, asbestos, and mold exposure can lead to acute and chronic health concerns. Involving a public health official or respiratory specialist can help the ISO develop mitigation strategies to prevent associated ailments.

In addition to general duties, the ISO/ASO-RT should consider specific evaluations and actions that can improve responder safety, including technical assistance, air monitoring, and improvisation monitoring.

Technical Assistance   Experience has repeatedly shown that consultation with a structural engineer pays dividends. Some departments have retained structural engineers for emergency call-in. Large cities may have structural engineer employees in the building department. Regardless of their origin, structural engineers provide invaluable assistance. Drywall/concrete dust, asbestos, and mold exposure can lead to acute and chronic health concerns. Involving a public health official or respiratory specialist can help the ISO develop mitigation strategies to prevent associated ailments. The fire department HSO can assist with exposure reports and documentation of the inevitable barrage of "nuisance" injuries (cuts, blisters, and contusions) that always seem to accompany collapse incidents.

Air Monitoring   Rescue technicians (and the ASO-RT) usually begin air monitoring early in their efforts. During basic and light collapse incidents, this essential task may be forgotten. Simple four-gas monitors can help responders become aware of oxygen levels and the presence of sewer gas. Natural and propane gas detectors are also useful.

Improvisation Monitoring   "Adapt and overcome" is a phrase embraced by firefighters. How talented firefighters improvise their way through a situation is truly amazing, and the building collapse incident often showcases these abilities. From the ISO perspective, improvisation should be continually evaluated to identify when responders are pushing the envelope. Use your building construction knowledge (imposition of loads, material characteristics, etc.).

Industrial Entrapment

The hazards and considerations associated with industrial accidents are as numerous as the types of processes and plants. Unusual machines, conveyors, bizarre chemicals, technologically advanced processes, and supersized equipment can present challenges and injury potential for responders. (See Hazards at Industrial Entrapments.)

Hazards at Industrial Entrapments

- Heavy machinery
- Complicated access
- Unsecured hazardous energy
- Hazmat
- Noise
- Interfaced and/or automated systems
- Security system impediment
- Megasized equipment
- Pinch hazards
- Equipment congestion
- Exotic materials
- Material stockpiling

In most industrial rescue cases, on-site employees shut down operating equipment that may further affect the victim. The ISO should double-check that lockout or tag-out measures have been implemented. Reliance on on-site expertise at the industrial plant presents a double-edged sword. Usually, nobody knows the equipment better than those who operate it, yet their

！ Safety

When evaluating rescue efforts, remember the basic law of motion: For every action, there is an equal and opposite reaction.

■ Note

As in all tech rescues, anyone performing high-concentration tasks needs an opportunity to take a mental (and physical) break.

comfort with the equipment may underestimate the hazards that the responder faces.

When evaluating rescue efforts, remember the basic law of motion: For every action, there is an equal and opposite reaction. Engineered components may be load stressed and spring out when cut. Rescue tools and equipment may be pushed beyond their designed limits; watch (and listen) for power equipment that is bogging down under load. As in all tech rescues, anyone performing high-concentration tasks needs an opportunity to take a mental (and physical) break.

Cave-Ins

The generic term "cave-in" can be applied to trench collapses, earthen slides (mud and rock), avalanches, and material entrapments (grain, sand, logs, etc.). Like other tech rescues, cave-ins come with their own set of hazards (see Hazards at Cave-Ins).

Hazards at Cave-Ins

- Shifting/unstable material
- Hidden infrastructure
- Oxygen deficiency
- Weather exposure
- Difficult slope or grade
- Poor footing
- Sink potential
- Secondary collapse
- Crush potential

Trench rescues require specific regulated procedures (29 CFR 1926-650). The ISO must develop a site safety plan, emergency procedures, and safety briefings. Using the LCES approach (*lookouts, communications, escape routes, and safety zones, see Chapter 14) for developing safety briefings is useful for all types of cave-ins and can remind the ISO and responders of essential safety elements:

- *Lookouts:* ASOs, soil engineers, and briefed support personnel can serve as lookouts. Binoculars and signal devices (like a whistle) can help them in this function.
- *Communications:* Visual and voice communications are often a viable communications tool. Do not forget to communicate the incident action plan to all responders as part of the safety briefing. Everyone

should know the "all-evac" signal that lookouts use should the unexpected occur.

- *Escape routes:* Escape ladders and boarded footpaths should be used to create reliable escape routes. Tethered rescuers may require mechanical or powered assistance for rapid escape. In these cases, technician-level rescue personnel should be used as the rescuer and escape assistant.
- *Safe zones:* Developing suitable safe zones can prove challenging. A separate shore or refuse area may need to be erected prior to victim extrication efforts. Natural and structural barriers can be explored.

Other unique hazards may require ISO attention. Exhaust fume accumulation, ground vibration, and specialized hydrovac equipment require awareness and attention (**Figure 16-4**). Finally, do not forget that one undeniable force: gravity.

■ Note
One would think that firefighters' intimate relationship with water would prepare them to understand its daunting force.

Water Rescues

One would think that firefighters' intimate relationship with water would prepare them to understand its daunting force. Yet perplexingly, the very

Figure 16-4 *A trench rescue operation requires that all necessary safety equipment and precautions be in place prior to rescuers entering the trench.*

ingredient that firefighters regularly use has caused numerous firefighter deaths. Water incidents can include swift water, lake, oceanic, flood, and ice situations. Each can present hazards for the rescuer (see Hazards at Water Incidents).

Hazards at Water Incidents

- Swift/hidden currents
- Low-head dams
- Submerged entrapment hazards
- Floating debris
- Electrocution
- Hypothermia
- Reduced visibility (murky water and water/salt spray)
- Fragile and/or shifting ice
- Marine life
- Frightened animals
- Distance to solid ground
- Crushing wave forces and undertows or riptides

Protection from elements and appropriate PPE issues commonly lead the list of ISO concerns at the water incident. Planning for rapid rescuer intervention should be also weighted heavily. Timekeeping can help the IC and ISO make judgments about rescue profiles and risk/benefit decisions. Dive-rescue certified responders can offer judgment regarding rescue profiles. Once a decision is made that a dive operation has changed from a rescue to a recovery, risk taking should be reduced (**Figure 16-5**).

Flood incidents can present multiple rescue situations when resources become quickly overtaxed, as evidenced by the 2005 Katrina hurricane and subsequent levy breaks in New Orleans. Responders to Katrina were faced with overwhelming difficulties in all phases of the rescue effort. From the numerous accounts of the Katrina response, you may glean some useful vicarious lessons that may help you perform ISO functions at a flood. Of particular interest is the numerous health issues that plague the response. Katrina responders reported working in a literal "cesspool" condition as the water receded.[1] Even small floods can create health hazards. Soliciting input from public health and environmental professionals can help the ISO address the concerns.

High-Angle Rescues

The popular trend of extreme sports has heightened fire department awareness of their ability (or inability) to rescue victims from radio/cell towers,

Figure 16-5 *Dive-rescue certified responders can offer judgment regarding the rescue profile of an incident. (Photo courtesy of Keith Muratori of FIREGROUND-IMAGES.COM.)*

elevated water storage tanks and bridge spans, high-rises, and precarious cliffs. Likewise, trapped or injured maintenance workers can find themselves in need of rescue from some amazingly challenging locations (**Figure 16-6**). High-angle rescues can present hazards that can chill even the most daring firefighter (see High-Angle Hazards).

High-Angle Hazards

- Limited access
- Dizzying heights
- Limited escape routes
- Slip/fall hazards
- Lightning/wind
- Limited anchor options
- Electrocution
- Heights beyond equipment capabilities
- Use of helicopters
- Equipment failure
- Falling debris
- Dropped equipment

Figure 16-6 *Thrill seekers and maintenance personnel injured or trapped in elevated places can present challenging rescues for fire departments.*

! **Safety**

When it looks like rescuers will be committed to a rope (or climb) for an extended period, prehydration and energy bar intake are warranted. If possible, suggest to the rescuers that they take a water bottle and an extra energy bar.

Risk/benefit evaluations are paramount when performing ISO functions at the high-angle incident. Secondarily, the ISO should ascertain whether the rescuers are trained and *willing* to engage in the operation. (While the fear of public speaking is the number one fear of most adults, the fear of falling from heights is number two.) Some firefighters freely admit that they are uncomfortable with dizzying heights; others may not be so forthcoming. While it sounds unconscionable that a firefighter would not be willing to work from heights, it is a reality. Thankfully, most firefighters can work through and overcome their fear—*most*. The ISO should watch for signs that a firefighter is losing concentration or becoming stressed by fear (rapid breathing, wide-eyes, and/or uncontrollable shaking).

Rescuers trained to hang from ropes typically double- and triple-check anchors and rigging as a matter of course. Utilizing an ASO-RT can add redundancy to these checks. When it looks like rescuers will be committed to a rope (or climb) for an extended period, prehydration and energy bar intake are warranted. If possible, suggest to the rescuers that they take a water bottle and an extra energy bar. While the ASO-RT monitors the rescue technicians, the ISO should monitor the actions and focus of support personnel and the surrounding environment. Understandably, support personnel (and onlookers) spend an inordinate amount of time looking up (or down, as the case may be) during the rescue. Other hazards may catch someone unsuspecting as they train their attention on the rescuers and victim; be their wingman.

Obtaining weather forecasts and watching the sky (see Chapter 9) can help the ISO be proactive in preparing rescuers for wind gusts, lightning, and precipitation. Nighttime operations present additional concerns. The introduction of artificial light, while logical, may actually cause further problems. As an example, rescuers climbing on a radio tower may experience night blindness as spotlights are trained at steep angles on their ascent. The use of glow sticks and soft headgear lighting by the climbers may be preferable on the ascent and descent. Spotlights can illuminate ascent and descent hazards in the path of the rescuers and/or illuminate the victim for reassurance and give rescuers a target point. The progress of a scaling rescuer can be monitored using a thermal imaging camera (TIC) in some cases.

The ISO also has to deal with the concerns resulting from the gathering of crowds and the media (see general ISO duties in the preceding section).

● Caution

Nighttime operations present additional concerns. The introduction of artificial light, while logical, may actually cause further problems.

Confined Spaces

As with hazmat and trench operations, confined space rescues are governed by OSHA regulations (29 CFR 1910.146). The use of an ISO/ASO-RT, the development of a site safety/emergency plan, and safety briefings are mandatory. Hazards that may require intervention at the confined space incident are listed in the accompanying box, Hazards at the Confined Space Rescue.

Hazards at the Confined Space Rescue

- Limited access/escape options
- Toxic/flammable atmospheres
- Oxygen deficiency
- Hazardous energy
- Communication difficulties
- Collapse
- Cramped quarters, limited mobility
- Distance that exceeds airlines, ropes, etc.
- Rust and mold, residues

Roadway/Transportation Incidents

Perhaps the most alarming trend in firefighter injuries and deaths is the increase in traffic-related incidents, and the increasing congestion in our communities is likely only to get worse. Firefighting departments must be compelled to improve response and safety control measures at roadway incidents. Rarely does an ISO respond (or get appointed) to a typical motor vehicle accident (MVA) unless the situation becomes a "working" incident with the need for more resources. Duty ISOs should consider a discretionary response when dispatch information paints a picture that warrants a response. Some examples are:

- Multiple vehicles involved
- Long response time
- Involvement of hazardous energy (power utilities, hydrants, bridges, and so on)
- Extreme weather
- Involvement of buses, hazmat, high-angle, and the like

In addition to roadway incidents, this section applies to railway, subway, and aircraft incidents. Each can present hazards that need to be evaluated by the ISO (see Hazards at Roadway/Transportation Incidents).

Hazards at Roadway/Transportation Incidents

- Other traffic or congestion
- Threat of nearby secondary crash
- Limited access or escape options
- Hazmat/munitions
- Fuels ignition/alternative fuels
- Damaged infrastructure
- Hazardous energy
- Heavy entanglement
- Weather exposure
- Instability
- Vehicle hazards (see Chapter 9)
- Bloodborne pathogens

Roadway Incidents   Saving lives is what firefighters are trained to do and most of them would risk their lives to save lives. However, the roadway incident should be an exception. Given the trends in traffic-related firefighter deaths,

Figure 16-7 *The number one safety consideration at roadway incidents is the threat of being hit by other traffic.*

⚠ Safety

Creating barriers, work zones, and traffic-calming processes are essential and should be the first priority for roadway incidents. The reality is stunning: We are at tremendous risk just investigating motor vehicle accidents.

responders must save their own lives first. The number one safety consideration at roadway incidents is the threat of being hit by other traffic (**Figure 16-7**). Creating barriers, work zones, and traffic-calming processes are essential and should be the first priority for roadway incidents. The reality is stunning: We are at tremendous risk just investigating motor vehicle accidents.

All other things being equal, the ISO at an MVA should focus more on surrounding elements (especially other traffic) than on the rescue itself. Certainly rescuers working near damaged pad transformers, downed wires, and rushing water are at risk and may need ISO attention, but if they are hit in a secondary crash, that is sure to complicate things. In the 1980s and 1990s, law enforcement personnel were reluctant to shut down traffic in support of MVA rescues. Fire officers from this era literally argued with law enforcement officers that traffic needed to be shut down. (Some of the fire officers were even arrested and went to jail!) With the increase in secondary crashes that have injured police and fire personnel, the reluctance seems to be ebbing away, although it may still be present in a few jurisdictions. Still, the ISO may want traffic diverted from the accident scene and must use well articulated facts to make that happen.

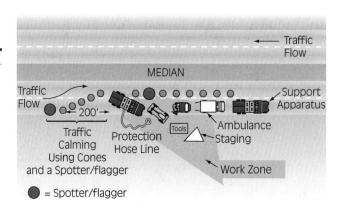

Figure 16-8 *The first-arriving large fire apparatus should be positioned to create a traffic barrier and work zone. Cones and a spotter/flagger can help with "traffic calming."*

traffic barrier
some object (like a large fire apparatus) that can absorb the impact of a secondary crash to protect rescuers; traffic barriers should be used to create a work zone, shielded from moving traffic, for rescuers

traffic-calming strategies
efforts to slow down approaching traffic: traffic cones, spotters or flaggers, arrow sticks, flashing lights, and warning signs

● **Caution**
Using a traffic barrier without traffic-calming strategies is *not* advisable. Worse, using traffic-calming strategies without a traffic barrier is downright dangerous.

As already mentioned, the basic strategy for improving rescuer safety at roadway incidents is to establish a system of traffic barriers, traffic-calming strategies, work areas, and escape zones (**Figure 16-8**). A **traffic barrier** can be defined as an object (like a large fire apparatus) that can absorb the impact of a secondary crash to protect rescuers. Remember "big fish eat little fish." When positioning the barrier (usually an apparatus that is not being utilized), make sure that, if hit, it is driven away from the rescuers and other traffic. The traffic barrier should create a work zone, shielded from moving traffic, for rescuers. **Traffic-calming strategies** include efforts to slow down approaching traffic: traffic cones, spotters or flaggers, arrow sticks, flashing lights, and warning signs. Using a traffic barrier without traffic-calming strategies is *not* advisable. Worse, using traffic-calming strategies without a traffic barrier is downright dangerous.

Another consideration for the ISO: At night, have apparatus operators minimize the use of white lights and strobes that can cause night blindness and increase the risk of a secondary crash.

Railway/Subway Incidents   The disaster drills that start off simple—only to have the facilitator throw in a hazmat, then a collapse, then an equipment malfunction, and then an explosion—may seem unrealistic. But it can all get real at the train derailment. The ISO must expect the worst to happen at a railway or subway incident. Imagine a subway train that has derailed between stops underground. In essence, you have a confined space, hazmat, industrial entrapment, and structural collapse—all rolled into one. Throw in a fire, and you have the ingredients of the most challenging incidents. As the ISO, you must become strategic and rely on ASOs to monitor rescuers. Use the considerations in relevant preceding sections to address the ISO functions at the subway (or railway) incident.

Aircraft Incidents Like the railway incident, an aircraft crash can present a variety of challenges. Using building collapse classifications (basic, light, moderate, heavy) can help the ISO judge the magnitude and potential hazards associated with the crash. The classification of the aircraft incident is influenced by the size of the aircraft and the size/type of the building that was hit. Regardless of the incident magnitude, a rescue profile needs to be established.

When the situation is viewed to be recovery in nature, the ISO should implement risk-reduction strategies. In the United States, representatives from the National Transportation Safety Board or Department of Defense (for military aircraft) respond and ask that the incident be treated like a crime scene. If this mind-set is taken and communicated before reps arrive, risk reduction can be achieved as working responders try to minimize the disruption and destruction of potential evidence.

Catastrophic crashes introduce the need for greater attention to the hazards of blood-borne pathogens. Gearing-up for this should be suggested by the ISO. Unburned jet fuel residue can be very damaging to protective equipment, and decontamination efforts may have to be implemented. Likewise, jet fuel vapors, burnt plastics, and composite metal dust are respiratory irritants and/or toxins—do not be quick to allow responders to doff SCBA during operations at aircraft incidents.

SUMMARY

Firefighters' "can-do" attitude compel them to action whether they are trained for the situation or not. This trap can create significant issues for the ISO at technical-rescue incidents. The history of injuries and deaths of rescuers at tech rescues has spawned the development of several regulations to hold rescuers accountable. Being familiar with these regulations can improve ISO performance. Many of the regulations require an ISO/ASO-RT to be present, as well as the development of safety plans and safety briefings. The general duties of an ISO are still applicable at the tech-rescue incident, and attention needs to be focused on the rescue/recovery profile and risk reduction. Rehab issues are focused on energy replacement and mental breaks to

help maximize the abilities of responders to concentrate.

A tech-rescue incident can be generally classified as collapse, industrial entrapment, confined space, roadway/transportation, water, or high-angle. Each of these categories presents unique hazards that the ISO needs to address. The use of LCES (*l*ookouts, *c*ommunications, *e*scape routes, and *s*afe zones) can help the ISO design a meaningful safety briefing for most tech-rescue incidents. At roadway incidents, the ISO should focus efforts on reducing the chance or impact of secondary crashes by the use of traffic barriers and traffic-calming strategies to create work zones. Railway and aircraft incidents can present numerous types of tech-rescue categories and challenge the ISO.

KEY TERMS

Assistant safety officer—rescue tech (ASO-RT) A person who meets or exceeds the NFPA 1670 requirements for Rescue Technician and is trained in the responsibilities of the ISO position as it relates to the specific rescue incident.

Basic/surface collapse Collapse in which victims are easily accessible and trapped by surface debris. Loads are minimal and easily moved by rescuers. The threat of secondary collapse is minimal.

Heavy collapse A collapse in which stressed concrete, reinforced concrete, and steel girders are impeding access. Included are collapses that require the response of USAR teams and specialized equipment and collapses that threaten other structures or that involve the possibility of significant secondary collapse.

Light collapse A collapse in which usually a light-frame (wood) partition collapses and common fire department equipment (from engine and truck companies) can access or shore areas for search and extrication. The threats of secondary collapse can be mitigated easily.

Moderate collapse A collapse of ordinary construction that involves masonry materials and heavy wood. Lightweight construction with unstable large open spans should also be classified as moderate. Significant void space concerns are present.

On-deck system An organized system in which a working team is replaced with another working team that is already dialed in and ready to replace them.

Traffic barrier Some object (like a large fire apparatus) that can absorb the impact of a secondary crash to protect rescuers. Traffic barriers should be used to create a work zone, shielded from moving traffic, for rescuers.

Traffic-calming strategies Efforts to slow down approaching traffic: traffic cones, spotters or flaggers, arrow sticks, flashing lights, and warning signs.

POINTS TO PONDER

Flash Floods and Storm Drains

A fire department was notified of several cars stranded due to heavy amounts of rain and subsequent flooding. A crew was dispatched to the scene at approximately 1700 hours to assist motorists stranded by the floodwaters. Upon arrival, the crew's captain sent two firefighters to check for any motorists in need of assistance. The two waded through approximately knee- to waist-high water for two blocks, checking several cars floating in the water. After approximately 15 minutes on the scene, the crew radioed their captain that all of the civilian motorists had exited their cars. After the crew determined that there were no civilians in the cars, they directed traffic away from the pooled water and waited until the police arrived to take over scene control. While the two firefighters were waiting
(continued)

(*continued*)

for the police to arrive, they were verbally summoned by a civilian bystander to help a female civilian stranded in the water. The civilian was observed holding onto a pole in a large pool of water. The crew believed the water was only 3 feet deep and that the civilian was standing on the ground. The civilian was actually standing on the top edge of a culvert on a large slope into the pool of water, which was approximately 10 feet deep.

Both of the firefighters responded to the location of the female civilian and attempted a rescue. The firefighters were wearing bunker pants, coats, boots, gloves, and helmets. Neither had received water rescue training. The first firefighter to enter the water was quickly pulled under by the undertow. The second firefighter then entered the water to aid his fellow firefighter. A grab was made and they both struggled to the edge of the water. The first firefighter, with his back to the water, climbed onto the bank, coughing from water he had swallowed. When the second firefighter reentered the water to assist the civilian, he told his partner to radio for help. As the first firefighter turned around, his partner was gone and his helmet was circling on the surface of the water. Seeing this, he removed his bunker coat and told civilian witnesses to use the radio and call for help. He reentered the water and assisted the civilian to safety. Witnesses found a welding cable and tied it around the firefighter's waist. He reentered the water and began to frantically search for his partner under the surface of the water. At 1744 hours, a female civilian witness used the radio to call for help. The captain, several hundred feet to the south of the scene, was confused about who was on the radio. Since his crew did not have any females on duty, he first thought it was a firefighter on another scene.

After an unsuccessful search for his partner, the firefighter exited the water and called on the radio for help, stating that a firefighter was down. The captain and another firefighter then ran to the location.

For several hours, dive crews and firefighters made numerous attempts to locate and rescue the victim. At approximately 2245 hours, the victim was found several blocks from the original location of the attempted rescue. He was pronounced dead at the scene.

The NIOSH investigation outlined several recommendations to prevent a reoccurrence. Some of the recommendations follow:

- Ensure that a proper scene size-up is conducted before performing any rescue operation, and that applicable information is relayed to the officer in charge.
- Ensure that all rescue personnel are provided and wear appropriate personal protective equipment when operating at a water incident.
- Ensure that firefighters who could potentially perform a water rescue are trained and utilize the "reach, throw, row, and go" technique.
- Develop site surveys for existing water hazards.
- Ensure that standard operating procedures (SOPs) are developed and utilized when water rescues are performed.

For Discussion:

1. What would you consider to be the prevailing factor contributing to this tragedy?
2. What personal safety system elements do you feel were missing (or incomplete)?
3. One recommendation is the provision of appropriate PPE for water rescues. For the first-due engine, what should that include?
4. Not all the preventive recommendations are listed here. What other types of preventive actions would you explore?

Note:

This case study was developed using NIOSH firefighter fatality investigative report 2001-02. Available at: www.cdc.gov/niosh/fire/reports/face200102.html.

REVIEW QUESTIONS

1. List several regulations that outline response requirements for tech-rescue incidents.

2. When should an ASO-RT be assigned at a tech-rescue incident?

3. For whom does the ASO-RT work at a tech-rescue incident?

4. List the two rehab issues that require special attention at tech-rescue incidents.

5. What is the benefit of using an on-deck system for crew rotation?

6. A collapsed building made of cinder block walls and small timber beams should be classified as which type of collapse?

7. List five hazards associated with industrial entrapments.

8. What is LCES and how can it be used at a cave-in?

9. List six hazards associated with water rescues.

10. List five hazards associated with high-angle rescues.

11. Name five circumstances in which a duty ISO should implement a discretionary response to motor vehicle accidents.

12. What is the difference between traffic barriers and traffic-calming strategies?

13. Diagram a strategic approach to protect rescuers at roadway incidents.

14. A subway derailment below ground is similar to what kind of incident?

15. What do jet fuel, burnt plastics, and composite metals have in common at an aircraft incident?

ADDITIONAL RESOURCES

Brown, Michael G. *Engineering Practical Rope Rescue Systems.* Clifton Park, NY: Delmar, a division of Thomson Learning, 2000.

Browne, George J., and Gus S. Crist. *Confined Space Rescue.* Clifton Park, NY: Delmar, a division of Thomson Learning, 1999.

Downey, Ray. *The Rescue Company.* Tulsa, OK: Fire Engineering Books and Video, a division of PennWell, 1992.

Linton, Steven, and Damon Rust. *Ice Rescue.* Fort Collins, CO: International Association of Dive Rescue Specialists, 1982.

NFPA 1006, *Standard on Professional Qualification for Rescue Technicians.* Quincy, MA: National Fire Protection Association, 2003.

NFPA 1670, *Standard on Operations and Training for Technical Search and Rescue Inci-* *dents.* Quincy, MA: National Fire Protection Association, 2004.

"Safety Management in Disaster and Terrorism Response." *Protecting Emergency Responders* Vol. 3, NIOSH Publication No. 2004-144. Available at: www.cdc.gov/niosh/docs/2004-144/chap7.html.

FA-159: *Technical Rescue Program Development Manual.* Emmitsburg, MD: United States Fire Administration, 1995.

NOTE

1. The author interviewed many Katrina responders who experienced disturbing health issues several weeks *after* their involvement.

Chapter 17

POSTINCIDENT RESPONSIBILITIES

Learning Objectives

Upon completion of this chapter, you should be able to:

- Explain the factors that lead to injuries during postincident operations.
- Discuss the role of the ISO for informal and formal postincident analysis (PIA).
- List the six specific items on which the ISO should provide input for a PIA.
- Explain the role of the ISO in accident investigation according to NFPA standards.
- List the five parts of the accident chain.
- List and explain the three steps of accident investigation.

INTRODUCTION

For every one serious firefighter injury, over six hundred near misses or close calls could easily have been serious.[1] At times, firefighters wear close calls as badges of courage and use exaggerated tales to fuel fire house coffee talk. Other times, firefighters involved with a near miss trivialize or minimize the brush with injury or death. Often, the closer firefighters come to serious injury, the more they minimize the storytelling, perhaps indicating that the event really got their attention. At what point does the incident safety officer need to follow up on a near miss and work toward the prevention of a similar event that may not have such "lucky" consequences?

■ **Note**
The lessons learned from *any* near miss should be folded into training and used for ongoing efforts to avoid similar situations in the future.

Ideally, the lessons learned from *any* near miss should be folded into training and used for ongoing efforts to avoid similar situations in the future. Often, the war stories that arise from close calls are invaluable tools in the teaching of new firefighters in academies nationwide. The key to making the lessons productive is an accurate portrayal of the facts and actions. Collecting information quickly and accurately can capture the information accurately. The reconnaissance responsibility places the ISO in the best position to collect and document incident activities for a postincident analysis or critique. Further, the ISO can use the information to begin an investigative process if an injury or fatality has occurred. This chapter explores the responsibilities and duties of the ISO for postincident activities, postincident analysis, and accident investigations.

POSTINCIDENT ACTIVITIES

■ **Note**
Postincident injuries seem ironic in a profession whose hallmark is aggressive and calculated risk taking.

While accurate documentation is spotty, many injuries seem to occur while crews are packing up to leave an incident. Common postincident injuries include strains, sprains, and being struck by objects. Postincident injuries seem ironic in a profession whose hallmark is aggressive and calculated risk taking. For each cause of postincident injuries, the ISO can take preventive steps to reduce their likelihood of happening.

Postincident Thought Patterns

postincident thought patterns
reflective or introspective mental wanderings that firefighters experience just after incident control

One cause of postincident injuries has to do with the little studied concept of postincident thought patterns. **Postincident thought patterns** are the reflective or introspective mental wanderings that firefighters experience just after incident control. These patterns can be summed up in one word: inattentiveness. In cases of especially difficult, unusually spectacular, or particularly challenging incidents, firefighters tend to reflect on their actions. The replay of the incident starts almost instantly when the order is given to "pick up."

Figure 17-1
Postincident introspection is normal but may lead to inattentiveness and injury.

Introspection is normal (**Figure 17-1**). The switch from activities requiring brainpower and physical effort to an activity that is so rudimentary as to be dull is a hard transition. Herein lies the problem.

Some incident commanders may release the ISO from the event after the "serious stuff" is done. This is an error. After a working incident, the ISO should circulate among those involved in the pick-up and keep an eye out for inattentiveness. Signs may include faraway stares or robotlike actions. Firefighters might stop and look about as if they have forgotten their task. Simple reminders or jocularity can help them regain focus and reduce their injury potential. One method to reduce the impact of postincident thought patterns is to take a brief time-out and have everyone gather for a quick incident summary and safety reminder. These huddles can be effective for everyone or just small groups (**Figure 17-2**). Even a casual coachlike approach that emphasizes the need to stay alert and not fall into an injury trap can be useful.

Chemical Imbalance

A successful rehab program prevents firefighters from experiencing fatigue and mental drain. Nevertheless, some may slip through the cracks, or many experience fatigue when rehab efforts are not effective. Even with effective rehab, the end of an incident, especially one requiring significant physical output, can cause chemical imbalance. With the end of an incident comes the relaxation of the firefighters' minds and the shutoff of protective chemicals that stimulate performance. The adrenaline rush is over and the firefighters' metabolisms return to a "repair" state; this reaction also causes a mental slowdown that can lead to unclear thinking and injuries.

■ Note
Simple reminders or jocularity can help them regain focus and reduce their injury potential.

Safety
One method to reduce the impact of postincident thought patterns is take a brief time-out and have everyone gather for a quick incident summary and safety reminder.

Figure 17-2 *Calling a huddle before incident cleanup is the best opportunity to remind firefighters of lingering injury threats.*

Another way to see the chemical and mind imbalance is to look at a firefighter's tools from a layperson's point of view. Laypersons literally have to concentrate on carrying an axe, pike-pole, chainsaw, roof ladder, or hose length in order to not cause harm to themselves or anyone around them. Put them in bulky, restrictive clothing and heavy boots, and you can see that they would probably get hurt doing almost anything. The firefighter does these things after incredible energy bursts under frightening conditions. Familiarity can play a certain role, but concentration is still required. If the mind has been taxed, the body has been fatigued and sends out the signal to relax. Yet the concentration requirement remains the same and the potential for injury rises.

Whether the issue is chemical imbalance or postincident thought patterns, the ISO needs to stay alert, pick up signs of potential injury, and take steps to "awaken" crews.

postincident analysis (PIA)

formal and/or informal reflective discussions that fire departments use to summarize the successes and improvement areas discovered from an incident

POSTINCIDENT ANALYSIS (PIA)

"Tailgate talk," "after-action report," "critique," "slam session," "incident review," and "Monday morning quarterbacking" are all labels the fire service has attached to the postincident analysis. The **postincident analysis (PIA)** is a formal and/or informal reflective discussion that fire departments use to summarize the successes and improvement areas discovered from an incident.

Successful fire officers always learn something from every working incident they are involved in. Each and every firefighter involved in an incident has a viewpoint or an opinion regarding specific circumstances or how an operation went, and these are important. The incident commander and the ISO bring perspectives to the incident overview. From this perspective the ISO should contribute—officially—to the postincident analysis.

NFPAs 1500 and 1521 require the ISO to be involved with the PIA. NFPA 1521 further stipulates that the ISO shall prepare a written report of pertinent information relative to health and safety issues.

To maximize the effect of safety-related input on a postincident analysis, the ISO needs to understand the essential philosophy of postincident analysis as well as ISO issues surrounding PIAs. This section contains a simple process to ensure that the ISO covers the appropriate information for the PIA.

PIA Philosophy

The ISO should approach any formal or informal postincident analysis with an attitude of positive reinforcement for safe habits and an honest, open desire to prevent future injuries. In most cases, the postincident analysis is nothing more than a discussion of what went right and what should be different next time. While this may sound simple, it is often hard to achieve, especially in light of a close call or a significant operational mistake that could have easily led to an injury.

When an operational mistake has been made, the ISO should first consider the likelihood and severity of an injurious outcome. If the possible effect warrants investigation, the ISO should employ a philosophy of discovery, which the ISO can approach from a fact-finding point of view. By asking a few questions about the operational environment or about the general feeling, or "pulse," of the incident, the ISO may help crews give voice to their concern or acknowledge the error.

Occasionally, this approach does not work. Crews may spend great amounts of energy "explaining" their actions in an attempt to justify them. In these cases, the ISO may then relate a personal account of the thought processes leading to a belief that crews might be injured. If this is accomplished with a communicated understanding of the crew's point of view and a sense of caring, a message will be sent. At all cost, avoid confrontation.

While the general approach of postincident analysis is to look *back* at an incident, the overriding goal is to look forward to the future. Rather than calling the postincident information "feedback," perhaps we should call it "feedforward."[2]

ISO PIA Issues

By monitoring an incident for potentially unsafe situations, the ISO brings many valuable observations to the postincident analysis. The subject areas addressed in the ISO Action Model (resources, recon, risk evaluation, and reporting) make it clear that the input from the ISO can be great and valuable. Nevertheless, a postincident analysis is a time for crews to share and reflect and take home a message. A long dissertation from the ISO can easily negate any such message. However, the ISO should comment on some key issues, including the following:

General Risk Profile of the Incident   The ISO can share the overall picture from a risk management point of view. Items such as risk/benefit, pace, and impressions about appropriateness of the risks taken can be discussed. If a situation developed that placed a crew at risk, the ISO may find it valuable to call on the crew to relay their thoughts or perceptions. These observations may have to be built on so that everyone takes away some value.

Effectiveness of Crew Tracking and Accountability   The ISO can yield to an accountability system manager for some of this type of information. Observations about crew freelancing (working in conflict with the action plan), individual freelancing (working without a partner), and reinforcement about successful tracking should be shared.

Rehabilitation Effectiveness   Even though they should, ISOs are seldom "processed" through rehab. How then can the ISO comment on the effectiveness of rehabilitation? The ISO can share observations of the pace, energy, and focus trends throughout the incident, as well as the duration or rotation of work efforts. If injuries resulted during the incident, an investigation is likely. But for postincident purposes, some exploration of rehab as a contributing factor may be discussed.

Personal Protective Equipment Use   Although the ISO normally shuttles individual PPE concerns to the company officer or crew leader, ongoing PPE issues can be addressed by the ISO. As an example, the choice to do overhaul at an incident without SCBA may have been premature. Likewise, a four-gas monitor may have been used to make the decision to go "packs-off" and use simple dust masks. These decisions can be discussed and reinforced as appropriate.

Close Calls   Obviously, the circumstances surrounding a near injury should be detailed from all participants' point of view. The ISO should minimize his or her contributions to close call events and reserve judgment because an actual investigation may be warranted.

Injury Status   If no injuries have been reported, this should be reinforced. Most likely, good practice and procedure led to no injuries. If this belief is not shared by all, then the role of "luck" should be discussed; this can very well raise safety awareness for the next incident. If a firefighter injury required transporting someone to a medical facility, firefighters will want an update. The ISO should be cautious. Issues of medical confidentiality, investigation results, and other ramifications may limit the amount of information that can be shared. In these cases, the ISO should keep the discussion centered on the efforts underway to take care of the injured as well as an overview of the investigative process.

If a firefighter injury was significant, or if a firefighter fatality occurred, the ISO should use the PIA as a tool to listen to firefighters. Often, firefighters may appear to be blaming when, in reality, they are venting, displacing stress, or even grieving. These reactions are normal.

PIA Process

A postincident analysis can take on many forms. Individual department procedure should dictate the best process to ensure that the ISO's observations are captured. Some general guidelines, however, can assist the ISO in preparing to contribute to a PIA.

First, the ISO should ascertain from the incident commander if he or she wishes to host a formal or informal PIA. Most routine incidents can be analyzed in an informal way just prior to releasing crews from the scene or just after cleanup at the fire station. This is especially effective if no significant operational issues have been raised. For significant incidents, a formal PIA should be prepared, perhaps taking days or weeks to prepare. If the formal PIA takes more than a few days to prepare, the IC and ISO should discuss doing an interim PIA to capture responders' memories; just let them know that it is an interim PIA and that a more complete session will be held in the future. If a firefighter fatality is involved, a formal PIA may be delayed until the investigation is complete. Whether formal or informal, the ISO should employ a few simple steps to make each an opportunity to increase everyone's ability to make solid risk decisions and prevent future injuries. Let's take a look at these simple steps.

On-Scene   Before crews pick up to leave the scene, make it a point to check in and say a few words (**Figure 17-3**). Also take an opportunity to ask a few questions. A successful technique is to ask crews if they noticed any hazards that you, the ISO, may have missed. Another simple, caring question is to ask if everyone on the crew is feeling OK or has received any minor injuries that you should know about. A parting positive comment regarding the crew's effort should always be included.

Documentation   At a minimum, the ISO should document a quick summary of the hazard issues discussed with the IC or any crew (**Figure 17-4**). This should be included with a summary of building construction features, any

Figure 17-3 *The ISO should make a point of checking in with crews to get a sense of their perspective of the incident.*

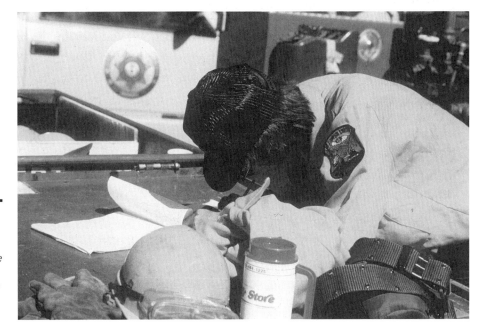

Figure 17-4 *Quick incident documentation is essential. The ISO perspective and documentation improve the quality of a postincident analysis.*

unique features of the operational environment, as well as an incident time line. A quick call to the dispatch center can help you record times. The chronological succession of events is often questioned during PIAs. With the oversight from the ISO, these questions can be clarified if the ISO has a documented time line.

If a formal PIA is scheduled, the ISO should spend more time with documentation and confer with the incident commander prior to the session. This avoids displaying differing points of view from the IC and ISO, which, in front of crews, can have the effect of "dissing" safety. Remember the forward focus as you prepare for a formal PIA.

Trend Spotting   The ISO who finds a recurring problem or concern should take the time to jot down some thoughts, then share them with a supervisor, training officer, or the department's health and safety officer. Remember: Many people can spot and articulate problems; only a few can present a problem along with some reasonable solutions.

■ Note

Often, the person to begin an accident investigation following a firefighter injury, fatality, or equipment mishap is the ISO, given the nature of his or her assignment as a command staff member.

NFPA

NFPA 1521 outlines the duties of the ISO in regard to accident investigation (paraphrased):

The incident safety officer shall initiate the accident investigation procedures as required by the fire department.

In the event of a serious injury, fatality, or other potentially harmful occurrence, the incident safety officer shall request assistance from the health and safety officer.

ACCIDENT INVESTIGATION

Another postincident ISO duty is incipient accident investigation. Often, the person to begin an accident investigation following a firefighter injury, fatality, or equipment mishap is the ISO, given the nature of his or her assignment as a command staff member. NFPA 1521 outlines the duties of the ISO in regard to accident investigation (paraphrased):

The incident safety officer shall initiate the accident investigation procedures as required by the fire department.

In the event of a serious injury, fatality, or other potentially harmful occurrence, the incident safety officer shall request assistance from the health and safety officer.

For significant (or multiple) injuries and deaths, the ISO should request HSO assistance for a simple reason: The role of the ISO is part of the equation that needs investigating. Outlining fire department procedures that need to be implemented following a fatality or an injury that may result in a fatality is beyond the scope of this book. This information is available through the United States Fire Administration, International Association of Fire Chiefs, and the International Association of Firefighters. In this section, we focus on the ISO role in accident investigations.

An accident investigation is one of the first steps in avoiding future injuries and deaths. Often, the results of the investigation can lead to the change of unsafe situations, habits, or equipment not only for the originating department, but for departments nationwide. This vicarious learning is essential. The Hackensack, New Jersey, tragedy is a perfect example.

In Hackensack, the local fire department was attempting to extinguish an advanced fire in the truss space (cockloft) of an automobile repair shop. The roof collapsed, trapping and eventually killing five firefighters. One accident report cited numerous contributing factors, including a failure to recognize the danger of collapse of truss-involved fires, command dysfunction, and a failure to have a firefighter accountability system. These findings became the impetus for change for fire departments around the country.

Close calls or near misses should also be investigated. Technically speaking, the phrases "close call" and "near miss" are interchangeable. Anecdotally, the phrase "near miss" can be regarded as a *near hit*. Regardless, we must learn from these events to prevent future injuries. Although the notion of a close call or near miss is subjective, it can be loosely defined as an unintentional, unsafe occurrence that could have produced an injury, fatality, or property damage; only a fortunate break in the chain of events prevented the undesirable outcome.

An open, nonjudgmental attitude toward close calls can a help a fire department realize the many warning signs, situational occurrences, and contributing factors that precede an injury. Consider also, the widely recognized *accident triangle* concept (**Figure 17-5**). The **accident triangle** is a statistics-based concept that, for every one serious injury, there are thirty minor injuries and over six hundred near misses or close calls. Most fire officers would probably agree that the accident triangle concept is accurate. Historically, the fire service has not documented close call events to the degree that other high-risk professions have (i.e., military, aeronautics, maritime). Frankly, this should scare all of us and motivate us to take lessons from so-called "practice accidents." Thankfully, a National Firefighter Near-Miss Reporting System has been created to help address this historical shortfall.[3] This system, administered by the International Association of Fire Chiefs, provides a voluntary, confidential, and secure reporting system with the goal of improving firefighter safety.

accident triangle
a statistics-based concept that, for every one serious injury, there are thirty minor injuries, and over six hundred near misses or close calls

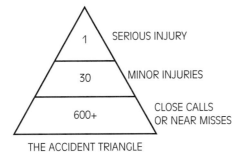

Figure 17-5   *The accident triangle shows that, for every one serious injury, there are thirty minor injuries and more than six hundred close calls. Close calls should also be investigated.*

1   SERIOUS INJURY

30   MINOR INJURIES

600+   CLOSE CALLS OR NEAR MISSES

THE ACCIDENT TRIANGLE

It is easy to see the benefit and importance of a thorough accident investigation. The well prepared ISO can truly make a difference in injury reduction through competent investigative skill.

Introduction to Accident Investigation

Accidents are the result of a series of conditions and events that lead to an unsafe situation that resulted in injury and/or property damage. Many call this series of conditions and events the *accident chain*. The investigation of an accident is actually the discovery and evaluation of the accident chain, which has five components (**Figure 17-6**):

- *Environment:* The physical surroundings, such as weather, surface conditions, access, lighting, and barriers
- *Human factors:* The components of human (or social) behavior—training, the use of/or failure to use recognized practices and procedures, fatigue, fitness, and attitudes
- *Equipment:* Personal protective equipment, its limitations and restrictions of equipment, its maintenance and serviceability, the appropriateness of its application, and, some may argue, the misuse of equipment (a human factor)
- *Event:* The intersection of the first three accident chain components, something to bring them together in such a way as to create an unsafe or unfavorable condition
- *Injury:* The injury or property damage associated with the accident (Because a near miss or close call is an accident without physical injury, for the sake of the accident chain, the injury can be supposed.)

THE ACCIDENT CHAIN

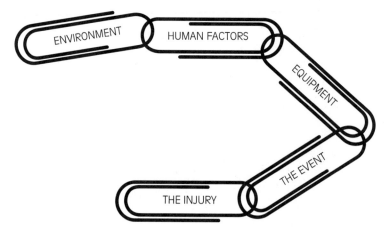

Figure 17-6 *Accident investigation is actually the discovery and linking of the accident chain.*

Ideally, the ISO should stop a potential accident by eliminating one of the elements in the chain *during* the incident. That is the focus of the ISO Action Model.

Investigation Issues

The ISO should be aware of issues and concerns that arise regarding involvement in accident investigation, one of the biggest of which is liability.

Picture this: A firefighter is seriously injured while working a commercial structure fire where an ISO is functioning (**Figure 17-7**). In the ensuing investigation, the question arises that, if a safety officer was present, should an injury have occurred? The conclusion of some may be that the ISO obviously did not do his or her job; it may lead to doubts about the objectivity of the ISO in investigating the incident. Did the ISO cover up anything in an attempt to deflect blame? This scenario may seem far-fetched, but it is a reality in our litigious society. How does the ISO perform safety tasks on-scene *and* conduct an honest, meaningful investigation following an accident? The answer is simple: Do both with due diligence. **Due diligence** is a legal phrase for the effort to act in a reasonable or prudent way, given the circumstances, with due regard to laws, standards, and accepted professional conduct.

The ISO who acts in a prudent manner—uses the ISO Action Model, takes steps to eliminate or communicate hazards, and works within established standards (NFPA) and laws (OSHA CFRs)—has taken significant steps in reducing liability. Added to this is a long-standing legal principle of *discretionary function*, which recognizes that certain activities require a

■ Note

Due diligence is a legal phrase for the effort to act in a reasonable or prudent way, given the circumstances, with due regard to laws, standards, and accepted professional conduct.

due diligence
a legal phrase for the effort to act in a reasonable or prudent way, given the circumstances, with due regard to laws, standards, and accepted professional conduct

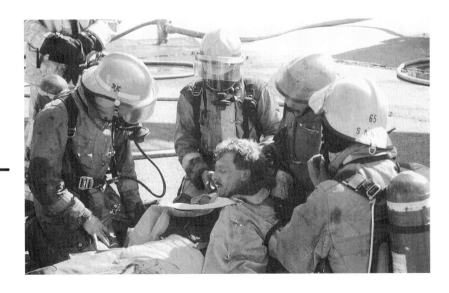

Figure 17-7 *The occurrence of a serious injury on the incident scene presents the ISO with many issues and concerns.*

value judgment among competing goals and priorities. In these cases, nonliability exists (*Nearing v. Weaver*).[4]

Another issue the ISO must be aware of in accident investigation is the involvement of outside agencies with an interest in the accident. State and/or federal OSHA and/or NIOSH officials, labor group investigators, insurance investigators, and law enforcement officials are often involved in a significant injury or death investigation. In many cases, these agencies can help the ISO; most likely, however, an investigation that has reached this magnitude signals the end of the ISO's need to lead or even participate in the investigation (the ISO becomes a witness).

■ **Note**

State and/or federal OSHA and/or NIOSH officials, labor group investigators, insurance investigators, and law enforcement officials are often involved in a significant injury or death investigation.

The Investigative Process

Where does the ISO begin to investigate an injury or mishap? Of the many investigative models to choose from, the most common is a simple three-step approach.[5]

Step 1: Information Collection   Numerous sources of information should be collected following an incident. These can be divided into six categories:

1. *Incident data:* Included is factual information, such as incident number, chronological time of events, weather conditions, apparatus assigned, personnel assigned, and documented benchmarks (primary search complete, incident under control).

2. *Witness statements:* These may be difficult to gather, and assistance may be required (law enforcement officials can usually help). An attempt is made to gain as many perspectives as possible (**Figure 17-8**). While keeping the witness speaking in facts is important, so is gaining a sense of the witness's perspective. Remember that much of what a firefighter does requires rapid judgment and execution.

3. *Scene sketches/diagrams:* Accuracy is key; be as precise as possible. Quick hand sketches work well for apparatus, hose, and crew placement, as long as measured distances are included so that a more precise drawing can be rendered later.

4. *Photographs/video:* If you noticed video footage being taken during an incident, attempt to gain it from the videographer. Media sources may be helpful. Follow-up video or still photography can capture the results of the accident.

5. *Physical evidence:* Protective equipment, damaged equipment, or other physical forms of evidence should be retained (**Figure 17-9**). Once again, law enforcement officials are a good source of expertise in the collection and documentation of physical evidence.

Figure 17-8 *The ISO should support an accident investigation with many witness reports.*

Figure 17-9 *Protective equipment and other physical evidence needs to be retained, marked, tagged, and identified.*

Figure 17-10 *Years of documentation may have to be reviewed to reconstruct an accident cause.*

6. *Existing records:* Equipment maintenance records, policy and procedure manuals, training records, and other documents are useful when it comes time to analyze the factors leading to the accident (**Figure 17-10**). At times, the extensive search may require going back many years to make the discovery that led to the injury. Likewise, the research may reveal that the proper maintenance, training, and other conditions were in place.

Step 2: Analysis and Reconstruction   The ISO reads through the accumulated data and separates facts, perceptions, unknowns, and determines the need for more information. At times, irrelevant data can be discarded. Once the information is analyzed, the ISO can reconstruct the accident. The reconstruction should be based on known facts, yet acknowledging the judgments and perceptions of involved crews. Once this is done, the accident chain should become apparent. Given the accident chain, the ISO can move to the next step.

■ Note

There may be a tendency to focus on one solution. Be inventive and force yourself to develop more than one solution.

Step 3: Recommendations   The chain of conditions and events that led to the accident indicates areas of concern. Most often, such concerns fall into the areas of equipment, policy and procedure, or personnel (training, attitude, fitness). From this point, the ISO can develop possible solutions to prevent the accident from happening again. There may be a tendency to focus on one solution. Be inventive and force yourself to develop more than one solution.

● **Caution**

Placing blame or recommending discipline has a tendency to close minds and erect barriers. If the ISO is to remain effective, it is best to state recommendations in the form of future accident prevention.

After multiple solutions are developed, evaluate each and focus on the approach that you believe would best prevent a future reoccurrence.

Nowhere in the accident investigative process is the word "blame" or "discipline." This is important. Placing blame or recommending discipline has a tendency to close minds and erect barriers. If the ISO is to remain effective, it is best to state recommendations in the form of future accident prevention. Upon discovering a case of complete disregard for safe practice (a form of negligence), the ISO should meet with a supervisor or chief officer and allow the department to handle the issue administratively. Although there may be some backlash and tension regarding the ISO's investigative effort, most safety conscious firefighters and officers will applaud the actions of the department and the ISO.

Accident investigation is not a fun task, yet it is vitally important to the reduction of future injuries. If the ISO can focus on good intent, the investigation serves as an investment in making a difference.

SUMMARY

The ISO plays a key role in preventing injuries during an incident, but, as crews begin to pick up, the injury potential persists. This threat is compounded by post incident thought patterns (introspection) and the possibility of chemical imbalance that can affect concentration. The ISO needs to remind firefighters of these phenomena to prevent postincident injuries.

The ISO is also in a key position to provide information for postincident analysis reports.

Specifically, the ISO should report on the safety points witnessed during the incident.

Following an incident injury, the ISO may be faced with serious issues regarding liability and blame. By exercising due diligence, the ISO can avoid such accusations. Regardless, the ISO must begin an accident investigation. By understanding the accident chain and applying a three-step investigative process, the ISO can discover and reconstruct the accident chain. With this information, the ISO can help prevent future injuries.

KEY TERMS

Accident triangle   A statistics-based concept that, for every one serious injury, there are thirty minor injuries, and over six hundred near misses or close calls.

Due diligence   A legal phrase for the effort to act in a reasonable or prudent way, given the circumstances, with due regard to laws, standards, and accepted professional conduct.

Postincident analysis (PIA)   Formal and/or informal reflective discussions that fire departments use to summarize the successes and improvement areas discovered from an incident.

Postincident thought patterns   Reflective or introspective mental wanderings that firefighters experience just after incident control.

POINTS TO PONDER

The Informal PIA

Firefighters Thompson and Leon were assigned to secure utilities at a single-family residential fire. After shutting off the natural gas feed, they proceeded to the electrical service. The breaker box was locked; after some discussion, they decided they could access the meter and shut off power that way. While pulling the meter, Firefighter Thompson felt a slight sting, accompanied by a brief electrical arc flash. Firefighter Thompson shook it off and finished his assignment.

As was customary for their fire department, all the firefighters involved in the house fire met for a brief informal postincident analysis. As each firefighter summarized their actions, it became apparent that Firefighters Thompson and Leon experienced a close call. The incident commander, Battalion Chief Sears, commented that pulling an electrical meter is quite dangerous and is not usually practiced in this fire department. The ensuing discussion revealed that the regional recruit academy teaches firefighters how to pull a meter and that individual local fire department policy should dictate whether this practice is allowed. Unfortunately, no written policy existed for Battalion Chief Sears' fire department. The assigned incident safety officer, Captain Drowns, was busy taking notes during the discussion and decided to reserve judgment.

For Discussion:

1. What type of action would you suggest to evaluate whether Firefighter Thompson was actually injured or not?
2. What type of questions should Captain Drowns ask of Firefighters Thompson and Leon during their informal PIA?
3. What issues do you see arising from Chief Sears' comment?
4. As an ISO, would you have reserved judgment in this case? Why?
5. Do you think a formal investigation is warranted for this close call?
6. What type of follow-up would you like to pursue in this case?

REVIEW QUESTIONS

1. What is meant by postincident thought patterns?

2. Compare and contrast the role of the ISO in informal and formal postincident analysis.

3. What are the six items that the ISO should comment on during a PIA?

4. Explain the role of the ISO in accident investigation according to NFPA standards.

5. What are the five parts of the accident chain?

6. Give examples of human factors in the accident chain.

7. What are the three steps of accident investigation?

8. What are the six pieces of information that should be collected for an accident investigation?

9. How should a complete disregard for safety be handled during the recommendation phase of an accident investigation?

ADDITIONAL RESOURCES

National Fallen Firefighters Foundation. Line-of-duty death procedures, support, and benefit information can be found through links to many such resources: Available at: http://www.firehero.net.

National Firefighter Near-Miss Reporting System. Firefighter near-miss reports and reporting procedures: Available at: http:// firefighternearmiss.com

Sovick, Mary. "Importance of Workplace Writing Skills in Fire and EMS Services." *The Instructor* (April 2004).

NOTES

1. Dennis G. Jones, "Accident Investigation Analysis," *Health & Safety for Fire and Emergency Service Personnel* 8, no. 9 (September 1997).

2. I first heard the feed-forward concept from coinstructor Terry Vavra, Deputy Chief, Lisle-Woodridge Fire District (Illinois), during one of our safety officer training sessions. I love the attitude it implies.

3. The National Firefighter Near-Miss Reporting System is funded through grants from the United States Fire Administration and the Fireman's Fund Insurance Company, and it is endorsed by the IAFC, IAFF, and the Volunteer and Combination Officers Section of the IAFC.

4. *Nearing v. Weaver*, 295 Ore. 702, 670 P.2d 137, 143 (Ore 1983). Timothy Callahan and Charles W. Bahme, *Fire Service and the Law* (Quincy, MA: National Fire Protection Association, 1987).

5. Ronald H. Hopkins, "Accident Investigation," *Health & Safety for Fire and Emergency Service Personnel* 5, no. 9 (September 1994).

Appendix

A

SAMPLE INCIDENT SAFETY OFFICER SOP

(Standard Operating Procedure): Selection, Training, and Duties

Purpose

To develop a system that provides a dedicated, on-call fire officer to respond to designated and significant incidents in order to serve the incident commander as the incident safety officer (ISO).

Responsibility

It shall be the fire and rescue department's responsibility to select and train eligible officers to serve as incident safety officers (ISOs). It shall be the selected fire officer's responsibility to adhere to the training and procedural requirements of this document.

PROCEDURE

General Program Guidelines

1. The Department shall select and train five to eight fire officers (career and volunteer mix) to serve as available incident safety officers. One career officer from each shift and three to four volunteers constitute an ideal and effective mix.

2. To be eligible for selection, a career or volunteer member must have at least five years of fire service experience, with at least two years' experience as a line officer (lieutenant or above) and be currently serving as an officer.

3. Serving as an ISO is voluntary. Members wishing to be selected as ISOs agree to complete the training program, attend required proficiency training, and serve at least two years as ISOs.

4. Selected officers who complete the ISO training shall design a system to ensure that one of the designated incident safety officers is available at all times. As a guideline, an off-duty career member trained as an ISO should be available using the same guidelines as the shift recall system.

5. Career members that are ISO-trained shall not serve as an ISO while on duty for a scheduled shift position, unless no other ISO is available and the incident commander makes this assignment.

6. ISOs completing the initial training program will be provided a safety officer vest, portable radio (career members), and clipboard/metal file. If the member no longer serves as an ISO, changes from officer status, or fails to complete training requirements, the member shall return ISO equipment to the department.

Training Requirements

1. All ISOs shall complete a 16-hour initial training session.

2. ISOs shall attend eight of twelve proficiency training sessions offered in conjunction with safety committee meetings.

3. Operating as an ISO at department live-fire training sessions, disaster drills, and full-scale HazMat scenarios shall be allowed to fill half of the annual proficiency training requirement.

4. ISOs shall pursue additional training in topic-specific areas, such as building construction, rehab and firefighter performance, tactics and strategies, fire behavior, and accident investigation.

Duties and Responsibilities

1. The ISO shall maintain all assigned equipment and ensure that he or she has the appropriate forms, checklists, and other tools prior to responding to incidents as the ISO.

2. The ISOs are responsible to establish and maintain a duty system that ensures that a safety officer will be available for response to incidents.

3. When responding to an incident, the duty ISO shall use his or her normal radio designation (C/O, F/C, chief, etc.). Upon arrival at an incident, the duty ISO shall make face-to-face contact with the IC, who shall confirm the ISO assignment.

4. The ISO shall be responsible for incident duties outlined in SOP #4-15, Use of Safety Officer at Incidents.

5. The ISO shall ensure that applicable follow-up care, investigation, and documentation are completed after incident scene injuries.

6. The ISO shall initiate applicable procedures for a firefighter line-of-duty death and be the lead officer in processing the PSOB application.

7. The ISO shall work with the IC and the department training officer in developing a postincident critique and appropriate recommendations for follow-up training.

Appendix B

SAMPLE INCIDENT SAFETY OFFICER SOP

(Standard Operating Procedure) Use of Safety Officer at Incidents

PURPOSE

To provide guidance and procedure for the utilization of an incident safety officer (ISO) at designated and significant incidents.

RESPONSIBILITY

It shall be the responsibility of incident commanders to designate an incident safety officer (ISO) at incidents outlined in this SOP. It shall be the responsibility of designated ISOs to follow procedure contained here within. It shall be the responsibility of all on-scene fire personnel to work with the ISO to recognize and minimize risks associated with incident environments and operations.

PROCEDURE

General Guidelines

1. The department recognizes that certain incidents present a significant or increased risk to firefighters. With these incidents come an increased responsibility to monitor firefighting actions and environmental conditions. The appointment of an incident safety officer can increase an incident commander's effectiveness in protecting firefighters.

2. The incident commander shall appoint an ISO early during an incident to maximize the effectiveness of the IC/ISO team. The ISO shall don a high-visibility safety officer vest to signify to all personnel the presence of the safety officer. Supervisors and leaders shall report hazards to the ISO in the course of operations.

3. An appointed incident safety officer shall have the authority of the incident commander to stop or alter any operation, action, or personal exposure that presents an imminent threat to the life safety of a firefighter, crew, or liaison person.

4. Any changed, altered, or stopped assignment made by the ISO shall be immediately communicated to the incident commander.

5. The ISO shall have the authority to appoint assistant safety officers if the size, scope, and duration of the incident warrant additional assistance.

6. The ISO shall possess a two-way radio and monitor radio transmissions. Imminent hazards shall be communicated to the incident commander and affected crews upon discovery.

Automatic ISO Response

The following types of dispatched incidents are cause for the duty ISO to respond to the incident:

1. Aircraft Alerts 2 and 3 involving index B or larger aircraft; also, any aircraft incident outside the airport critical zones

2. A reported fire in a commercial structure; including reports of heavy smoke or fire from a commercial structure, but *not* including odor investigations, automatic fire alarms, and other "investigative-type" alarms

3. Activation of a specialty team for an incident, including rope rescue, heavy rescue, dive rescue, hazmat, and wildland team activations

4. Incidents at "target" hazard complexes, including person trapped, collapses, smoke or odor reports, spills, and any other incident other than single patient EMS calls. Complexes include:

- Merix Corporation
- Walmart Distribution Center
- Omni-Trax
- Plaza Apartments
- Woodward-Governor
- Collins Plating
- Good Samaritan
- Colorado Crystal
- The Wexford
- McKee Medical Center

5. Any reported fire or working incident when climatic conditions become extreme, including:

- Temperatures over 97 degrees Fahrenheit
- Temperatures below 0 degrees Fahrenheit
- Winds gusting over 30 miles per hour
- Snow depths over 12 inches
- Wind-driven snow over 30 miles per hour

6. The duty ISO shall monitor incidents and self-dispatch if the communicated information suggests a difficult or significant incident (for example, an MVA with a school bus involved; explosion; hostage situation; etc.).

Automatic ISO Delegation

The incident commander shall delegate an officer to fill the ISO position if any of the following conditions exists:

1. A second (or greater) alarm is struck.

2. A firefighter injury requires transport or a line-of-duty death occurs.

3. Five or more group/division assignments or divisions of operations into branches are made.

4. Anytime firefighters must take extreme risks or the incident commander wishes to delegate the safety responsibility.

Specific ISO Duties

1. The duty ISO shall report to the IC and confirm the need for a safety officer prior to acting as one.

2. The ISO shall utilize the ISO checklist and prioritize safety efforts. Hazards found shall be communicated to the appropriate crew and the IC.

3. The ISO shall take the lead in the investigation and documentation of firefighter injuries on-scene.

4. When a significant safety violation is found, the ISO shall take action or cause to take action to ensure that the violation is addressed and follow-up is documented at the company level.

5. Upon hearing a Mayday, witnessing a collapse, or noting any other firefighter emergency condition, the ISO shall immediately report to the IC and assist with developing a systematic approach to the rescue and recovery of firefighters.

6. Before performing any reconnaissance or operation in the hot zone, the ISO shall team up with a partner and be tracked through the accountability system.

7. The ISO shall refrain from assisting with any task assignment and should maintain an observer role to best provide consultant information to the incident commander. If task involvement is required to prevent a firefighter injury, this action may take place; however, the task should be altered once the immediate threat is abated.

Appendix

C

SAMPLE ISO CHECKLISTS

Many checklists and forms have been developed to assist the ISO on scene. ISOs are encouraged to do some research and develop or adopt forms and checklists that can best work for their own jurisdictions or departments. Research options can include online sources, samples from other departments, and the examples included as Annex material in the NFPA 1521 standard.

In some cases, the ISO (or assigned ASOs) may be required to use a predesignated form.

For example, the ISO assigned to a Type 3 Wildland Fire will be required to use ICS Forms 214 and 215A.

Included in this appendix are a checklist developed to match the ISO Action Model and one developed for NIMS. Additionally, we've included some common ICS forms that an ISO may be required to use.

INCIDENT SAFETY OFFICER CHECKLIST

UPON ARRIVAL or ASSIGNMENT

- ☐ Don Safety Officer vest.
- ☐ Obtain briefing from Incident Commander, including:
  SITSTAT/RESTAT.
  Action plan.
  Known hazards or concerns.
- ☐ Prioritize Safety Officer duties:
  Risk review
  Reconnaissance
  Resource evaluation
  Reporting/planning
  (Each of these is listed below.)
- ☐ Start tracking incident elapsed time.
- ☐ Determine need for and request Assistant Safety Officers.

RISK REVIEW

- ☐ Define risk level for action plan:
  Life at risk
  Property at risk
  Stabilization only
- ☐ Determine frequency and severity of hazards.
- ☐ Prioritize hazard control recommendations.
- ☐ Address unacceptable risk situations:
  Stop or alter if life-threatening.
  Immediately advise IC.

RECONNAISSANCE—Environmental

- ☐ Perform 360-degree scene survey.
- ☐ Define principal hazard and its location.
  Validate control zones.
  READ SMOKE!
- ☐ Evaluate integrity of environment as:
  Stable—not likely to change.
  Starting to change.
  Rapidly changing.
- ☐ Classify structure(s) involved according to:
  Construction type.
  Materials used.
  Loads imposed.
  Confirm construction type and hazards with IC.
- ☐ Evaluate collapse potential, in terms of:
  Structural degradation.
  Time/duration that fire impacts building.
  Excessive loads.
  Scope of collapse.
  Stability profile after collapse.
- ☐ Define scope of utility involvement.
- ☐ Evaluate effects of weather on incident.
- ☐ Identify access/egress routes and deficiencies.
- ☐ Define traffic hazards.

RECONNAISSANCE—Operational

- ☐ Observe:
  Tactical assignments.
  Tactical effectiveness.
  Team effectiveness.
  Possible freelancing.
  Tool application and effectiveness.
  Action plan compatibility.
- ☐ Check the exposure of teams, in terms of:
  Correctness of PPE.
  Awareness of hazard.
  Appropriateness of risk level.
  Availability of escape routes.
- ☐ Determine the injury potential, especially:
  Fall/trip hazards.
- ☐ Evaluate the energy of crews (rehab profile).
- ☐ Evaluate apparatus placement and exposure.
- ☐ Monitor radio communications for:
  Dysfunction.
  Status reports/PARS.
  MAYDAY report.

RESOURCE EVALUATION

- ☐ Check scene attendance for:
  Too few/too many crews.
  Plans for additional crews.
- ☐ Determine rapid intervention status/capability.
- ☐ Determine the number of crews at risk.
- ☐ Check the effectiveness of the accountability system.
- ☐ Start time-pacing the incident:
  Use dispatch assistance
  Anticipate total on-scene time.
- ☐ Evaluate the rehab process and effectiveness.
- ☐ Assess the need for CISM.

REPORT/PLANNING

- ☐ Communicate concerns to the IC.
- ☐ Develop contingencies.
- ☐ Attend planning meetings.
- ☐ Review action plan revisions and updates.
- ☐ Prepare hazard awareness and safety briefings for arriving crews.

AFTER THE INCIDENT

- ☐ Prepare a report of still existing hazards.
- ☐ Document the Safety Officer actions.
- ☐ When ordered, secure operations.
- ☐ Provide information to the IC and Training Officer for incident critique.
- ☐ Complete accident/injury documentation IAW SOPs/SOGs.

SAFETY OFFICER POSITION CHECKLIST

The following checklist should be considered as the minimum requirements for this position. Note that some of the tasks are one-time actions; others are ongoing or repetitive for the duration of the incident.

✓	**Task**

☐   1. Obtain briefing from Incident Commander and/or from initial on-scene Safety Officer.

☐   2. Identify hazardous situations associated with the incident. Ensure adequate levels of protective equipment are available, and being used.

☐   3. Staff and organize function, as appropriate:
   - In multi-discipline incidents, consider the use of an Assistant Safety Officer from each discipline.
   - Multiple high-risk operations may require an Assistant Safety Officer at each site.
   - Request additional staff through incident chain of command.

☐   4. Identify potentially unsafe acts.

☐   5. Identify corrective actions and ensure implementation. Coordinate corrective action with Command and Operations.

☐   6. Ensure adequate sanitation and safety in food preparation.

☐   7. Debrief Assistant Safety Officers prior to Planning Meetings.

☐   8. Prepare Incident Action Plan Safety and Risk Analysis (USDA ICS Form 215A).

☐   9. Participate in Planning and Tactics Meetings:
   - Listen to tactical options being considered. If potentially unsafe, assist in identifying options, protective actions, or alternate tactics.
   - Discuss accidents/injuries to date. Make recommendations on preventative or corrective actions.

☐   10. Attend Planning meetings:

ICS 202

Sample Planning Meeting Agenda

Agenda Item	Responsible Party
1 Briefing on situation/resource status.	Planning/Operations Section Chiefs
2 Discuss safety issues.	Safety Officer
3 Set/confirm incident objectives.	Incident Commander
4 Plot control lines and Division boundaries.	Operations Section Chief
5 Specify tactics for each Division/Group.	Operations Section Chief
6 Specify resources needed for each Division/Group.	Operations/Planning Section Chiefs
7 Specify facilities and reporting locations.	Operations/Planning/Logistics Section Chiefs
8 Develop resource order.	Logistics Section Chief
9 Consider communications/medical/transportation plans.	Logistics/Planning Section Chiefs
10 Provide financial update.	Finance/Administration Section Chief
11 Discuss interagency liaison issues.	Liaison Officer
12 Discuss information issues.	Public Information Officer
13 Finalize/approve/implement plan.	Incident Commander/All

☐ 11. Participate in the development of Incident Action Plan (IAP):
   • Review and approve Medical Plan (ICS Form 206).
   • Provide Safety Message (ICS Form 202) and/or approved document.
   • Assist in the development of the "Special Instructions" block of ICS Form 204, as requested by the Planning Section.

☐ 12. Investigate accidents that have occurred within incident areas:
   • Ensure accident scene is preserved for investigation.
   • Ensure accident is properly documented.
   • Coordinate with incident Compensation and Claims Unit Leader, agency Risk Manager, and Occupational Safety and Health Administration (OSHA).
   • Prepare accident report as per agency policy, procedures, and direction.
   • Recommend corrective actions to Incident Commander and agency.

☐ 13. Coordinate critical incident stress, hazardous materials, and other debriefings, as necessary.

☐ 14. Document all activity on Unit Log (ICS Form 214).

INCIDENT OBJECTIVES	1. Incident Name	2. Date	3. Time
4. Operational Period			
5. General Control Objectives for the Incident (include alternatives)			
6. Weather Forecast for Period			
7. General Safety Message			

8.	Attachments (mark if attached)	
☐ Organization List—ICS 203	☐ Medical Plan—ICS 206	☐ (Other)
☐ Div. Assignment Lists—ICS 204	☐ Incident Map	☐
☐ Communications Plan—ICS 205	☐ Traffic Plan	☐
9. Prepared by (Planning Section Chief)	10. Approved by (Incident Commander)	

ICS 202

UNIT LOG	1. Incident Name	2. Date Prepared	3. Time Prepared
4. Unit Name/Designators	5. Unit Leader (Name and Position)		6. Operational Period

7. Personnel Roster Assigned

Name	ICS Position	Home Base

8. Activity Log

Time	Major Events

9. Prepared by (Name and Position)

INCIDENT ACTION PLAN SAFETY ANALYSIS

1. Incident Name	2. Date	3. Time

Division or Group	Potential Hazards	Mitigations (e.g. PPE, buddy system, escape routes)
Type of Hazard:		
Type of Hazard:		
Type of Hazard:		
Type of Hazard:		
Type of Hazard:		
Type of Hazard:		
Type of Hazard:		
Type of Hazard:		

Prepared by (Name and Position)

ICS 202

GLOSSARY

Accident triangle A statistics-based concept that, for every one serious injury, there are thirty minor injuries, and over six hundred near misses or close calls.

Action Model A template that outlines a mental or physical process to be followed.

Active cooling Using external methods or devices (such as hand and forearm immersion, misting fans, or ice vests) to reduce an elevated body core temperature.

Assistant safety officer—hazmat (ASO-HM) A person who meets or exceeds the NFPA 472 requirements for Hazardous Materials Technician and is trained in the responsibilities of the ISO position as it relates to hazmat response.

Assistant safety officer—rescue tech (ASO-RT) A person who meets or exceeds the NFPA 1670 requirements for Rescue Technician and is trained in the responsibilities of the ISO position as it relates to the specific rescue incident.

Basic/surface collapse A collapse in which victims are easily accessible and trapped by surface debris. Loads are minimal and easily moved by rescuers. The threat of secondary collapse is minimal.

Black fire A slang term used to describe high-volume, turbulent, ultradense, and deep-black smoke; a sure sign of impending flashover.

Blow-up A wildland fire term used to describe the sudden advancement and increase in fire intensity due to wind, prewarmed fuels, or topographic features, such as a narrow canyon or a chimney.

Case law Law that refers to a precedent (a ruling by a judge or a specific court proceeding) established over time through the judicial process.

Circadian rhythms A person's physiological response to the 24-hour clock, which includes sleep, energy peaks, and necessary body functions.

Code of Federal Regulations (CFRs) The body of laws enacted by OSHA that are used to help achieve workplace safety.

Cold Zone Establishes the public exclusion or clean zone. There are minimal risks for human injury and/or exposure in this zone. Denote cold zones with green tape.

Collapse zone Areas that are exposed to trauma, debris, and/or thrust of a collapse. A collapse zone is a more specific form of a no-entry zone.

Contamination reduction zone An area where decontamination takes place and includes a safe refuge area for contaminated victims and responders who have left (or who have rapidly escaped) the IDLH zone.

Due diligence A legal phrase for the effort to act in a reasonable or prudent way, given the circumstances, with due regard to laws, standards, and accepted professional conduct.

Engineered wood Products that consist of many pieces of native wood (chips, veneers, and saw dust) glued together to make a sheet, beam, or column.

Environmental integrity The state of a building, area, or condition being sound, whole, or intact.

Firm Intervention An intervention to immediately stop, alter, or suspend an action or operation due to an imminent threat; more or less an official order to stop, alter, or suspend an action.

FiRP Fiber-reinforced plastic (usually pronounced "ferp").

Flaring A sudden, short-lived rise in fire intensity, attributed to wind, fuel, or topographical changes. Flaring can be a warning sign of an upcoming blow-up.

Freelancing Failure to work within the framework of an incident action plan.

Ground gradient Electrical energy that has established a path to ground through the earth and that continues to energize the earth. A downed power line may be energizing the earth in a concentric ring of up to 30 feet, depending on the voltage of the source.

Guidelines Adaptable templates that give wide application flexibility.

Hazardous energy Stored potential energy that causes harm if suddenly released.

Health and safety officer (HSO) The member of a fire department assigned and authorized by the fire chief as the manager of the occupational safety and health program.

Heavy collapse A collapse in which stressed concrete, reinforced concrete, and steel girders are impeding access. Included are collapses that require the response of USAR teams and specialized equipment and collapses that threaten other structures or that involve the possibility of significant secondary collapse.

Hostile fire event An event that can catch firefighters off guard and endanger them: flashover, backdraft, smoke explosions, and rapid fire spread.

Hot zone The area presenting the greatest risk to members and will often be classified as an IDLH atmosphere. Denote hot zones with red tape.

Hybrid buildings Building construction methods that do not fit into the five classic building construction types. Hybrids also include buildings built using more than one type of method.

IDLH zone Areas in or around the building where working firefighters are exposed or may become exposed to smoke and heat. Persons working in an IDLH zone shall have a partner, work under the two-in/two-out rule, and be tracked through an accountability system.

IDLH An acronym given to environments that are immediately dangerous to life and health.

Imminent Threat An activity, condition, or inaction that will most certainly lead to injury or death.

Incident management team (IMT) A trained overhead IMS team with specific expertise and organized to deploy to incidents for management functions that exceed those available at the local level.

Incident safety officer (ISO) A member of the command staff responsible for monitoring and assessing safety hazards or unsafe situations and for developing measures for ensuring personnel safety at the scene of an incident.

ISO Action Model A cyclic, four-arena model that allows the incident safety officer to mentally process the surveying and monitoring functions of typical incident activities and concerns.

Laminar smoke flow The smooth and stable flow of smoke through a building; indicates that the building (or compartment) is still absorbing heat.

Laminated veneer lumber (LVL) Lumber created by gluing and pressing together sheet veneers of wood (in the same grain direction).

LCES An acronym that stands for *l*ookouts, *c*ommunication methods, *e*scape routes, and *s*afety zones.

Learning The acquisition of knowledge, skills, and attitude to achieve mastery.

Light collapse A collapse in which usually a light-frame (wood) partition collapses and common fire department equipment (from engine and truck companies) can access or shore areas for search and extrication. The threats of secondary collapse can be mitigated easily.

Mastery The concept that an individual can achieve 90 percent of an objective 90 percent of the time.

Moderate collapse A collapse of ordinary construction that involves masonry materials and heavy wood. Lightweight construction with unstable large open spans should also be classified as moderate. Significant void space concerns are present.

National Fire Protection Association (NFPA) A for-profit association recognized for developing consensus standards, guides, and codes for a whole realm of fire-related topics.

National Incident Management System (NIMS) A presidentially mandated, consistent nationwide approach to prepare, respond, and recover from domestic incidents regardless of cause, size, or complexity.

NIMS Integration Center (NIC) The center responsible for the ongoing development and refinement of various NIMS activities and programs.

No-entry zone Areas where no person—including firefighters, police, other responders, or the general public—should enter due to the serious or unpredictable nature of a hazard or condition. Denote no-entry zones with red/white chevron tape.

On-deck system An organized system in which a working team is replaced with another working team that is already dialed in and ready to replace them.

Oriented strand board (OSB) A wood sheeting consisting of wood chips (strands oriented in multiple directions) and an emulsified glue.

Personnel accountability report (PAR) An organized reporting activity designed to account for all personnel working an incident. To be truly effective, PAR radio transmissions should include assignment, location, and number of people in the assignment.

Passive cooling The use of shade, air movement, and rest to bring down human core temperatures.

Performance The demonstration of acquired mastery.

Postincident analysis (PIA) Formal and/or informal reflective discussions that fire departments use to summarize the successes and improvement areas discovered from an incident.

Postincident thought patterns Reflective or introspective mental wanderings that firefighters experience just after incident control.

Procedures Strict processes with little or no flexibility.

Reconnaissance (recon) An exploratory examination of the incident scene environment and operations.

Rescue profile A classification given to the probability that a victim will survive the environment. Typical classifications are high, moderate or zero for a given area in a building. In a zero rescue profile, there is obvious death or no chance for the victim to survive.

Risk management The process of minimizing the chance, degree, or probability of damage, loss, or injury.

Risk The chance of damage, injury, or loss.

Safety officer   According to NIMS, a member of the command staff responsible for monitoring and assessing safety hazards or unsafe situations and developing measures for ensuring personnel safety. *Note:* NFPA uses the title "incident safety officer" (ISO) for greater specificity.

SIP-wall (structural insulted panel)   A construction method that uses panels made from OSB and EPS for load-bearing walls and roofs. The panels are two sheets of OSB glued to both sides of an EPS sheet that is typically 6 to 8 inches thick.

Situational awareness   The degree of accuracy by which one's perception of the current environment mirrors reality. Applied to the ISO, situational awareness is the ability to accurately read potential risks and recognize factors that influence the incident outcome.

Smoke   The products of incomplete combustion, including an aggregate of solids, aerosols, and fire gases that are toxic, flammable, and volatile.

Soft Intervention   An intervention to make crews, command staff, and general staff aware that a hazard or injury potential exists.

Spalling   A pocket of concrete that has crumbled into fine particles through exposure to heat.

Statutory law   Laws promulgated to deal with civil and criminal matters.

Structural elements   Building columns, beams, and connections.

Support zone   Areas where firefighters, other responders, IMS staff, and apparatus are operating or staged. The general public should *not* be allowed to wander into the support zone.

Thermal protective performance (TPP)   A value rating given to the insulative quality of structural personal protective clothing and equipment.

Traffic barrier   Some object (like a large fire apparatus) that can absorb the impact of a secondary crash to protect rescuers. Traffic barriers should be used to create a work zone, shielded from moving traffic, for rescuers.

Traffic-calming strategies   Efforts to slow down approaching traffic: traffic cones, spotters or flaggers, arrow sticks, flashing lights, and warning signs.

Turbulent smoke flow   The movement of smoke through a building that is rapid and violent and that has expansive velocity (sometimes referred to as "agitated," "boiling," or "angry" smoke); indicates that the building (or compartment) cannot absorb more heat and is a precursor warning sign of flashover.

Valued property   Physical property whose loss will cause harm to the community.

Vicarious learning   Learning from the mistakes of others.

Warm Zone   A limited access area for members directly aiding or in support of operations in the hot zone. Denote warm zones with yellow tape.

ACRONYMS

ARFF Aircraft Rescue and Firefighting

ASO Assistant Safety Officer (or Assistant Incident Safety Officer)

ASO-HM Assistant Safety Officer—Hazmat

ASO-RT Assistant Safety Officer—Rescue Tech

CAD Computer Aided Dispatch

CFR Code of Federal Regulations

DHS Department of Homeland Security

EPS Expanded Polystyrene

FDSOA Fire Department Safety Officers Association

FEMA Federal Emergency Management Agency

FIRESCOPE *F*ire *R*esources of *S*outhern *C*alifornia *O*rganized for *P*otential *E*mergencies

FiRP Fiber reinforced plastic (usually pronounced "ferp")

HSO Health and Safety Officer

IAP Incident Action Plan

IC Incident Commander

ICF Insulated Concrete Forms

IDLH Immediately Dangerous to Life and Health

IMS Incident Management System

IMT Incident Management Team

ISO Incident Safety Officer

LCES *L*ookouts, *C*ommunication methods, *E*scape routes, and *S*afety zones.

LVL Laminated Veneer Lumber

MVA Motor Vehicle Accident

NFFF National Fallen Firefighters Foundation

NFPA National Fire Protection Association

NIC NIMS Integration Center

NIMS National Incident Management System

NIOSH National Institute for Occupational Safety and Health

NWCG National Wildfire Coordinating group

OSB Oriented Strand Board

OSHA Occupational Safety and Health Administration

PAR Personnel Accountability Report

PIA Postincident Analysis

PPE Personal Protective Equipment

PPV Positive Pressure Ventilation

RIC Rapid Intervention Crew/Company

SIP Structural Insulated Panels

SO Safety Officer

SOG Standard Operating Guideline

SOP Standard Operating Procedure

TIC Thermal Imaging Camera

TPP Thermal Protective Performance

USAR Urban Search and Rescue

USFA United States Fire Administration

WMD Weapons of Mass Destruction

INDEX